On the Animal Trail

Baptiste Morizot

On the Animal Trail

Translated by Andrew Brown

polity

Originally published in French as *Sur la piste animale* © Actes Sud, France, 2018

This English edition © Polity Press, 2021

Internal animal prints: Veronika Oliinyk/iStock

Polity Press
65 Bridge Street
Cambridge CB2 1UR, UK

Polity Press
101 Station Landing
Suite 300
Medford, MA 02155, USA

All rights reserved. Except for the quotation of short passages for the purpose of criticism and review, no part of this publication may be reproduced, stored in a retrieval system or transmitted, in any form or by any means, electronic, mechanical, photocopying, recording or otherwise, without the prior permission of the publisher.

ISBN-13: 978-1-5095-4717-3
ISBN-13: 978-1-5095-4718-0 (paperback)

A catalogue record for this book is available from the British Library.

Library of Congress Cataloging-in-Publication Data

Names: Morizot, Baptiste, author. | Brown, Andrew (Literary translator), translator.
Title: On the animal trail / Baptiste Morizot ; translated by Andrew Brown.
Other titles: Sur la piste animale. English.
Description: English edition | Cambridge ; Medford, MA : Polity, [2021] | "Originally published in French as Sur la piste animale © Actes Sud, France, 2018." | Includes bibliographical references. | Contents: Preamble: Enforesting oneself -- The signs of the wolf -- A single bear standing erect -- The patience of the panther -- The discreet art of tracking -- Lombric cosmology -- The origin of investigation. | Summary: "How paying attention to the tracks of animals can change our way of relating to the world around us"-- Provided by publisher. Identifiers: LCCN 2020054891 (print) | LCCN 2020054892 (ebook) | ISBN 9781509547173 (hardback) | ISBN 9781509547180 (paperback) | ISBN 9781509547197 (epub)
Subjects: LCSH: Human-animal relationships--Philosophy. | Tracking and trailing--Personal narratives.
Classification: LCC B105.A55 M6613 2021 (print) | LCC B105.A55 (ebook) | DDC 304.2/7--dc23 LC record available at https://lccn.loc.gov/2020054891
LC ebook record available at https://lccn.loc.gov/2020054892

Typeset in 11 on 14 pt Fournier MT by
Servis Filmsetting Ltd, Stockport, Cheshire
Printed and bound in Great Britain by TJ Books Ltd, Padstow, Cornwall

The publisher has used its best endeavours to ensure that the URLs for external websites referred to in this book are correct and active at the time of going to press. However, the publisher has no responsibility for the websites and can make no guarantee that a site will remain live or that the content is or will remain appropriate.

Every effort has been made to trace all copyright holders, but if any have been overlooked the publisher will be pleased to include any necessary credits in any subsequent reprint or edition.

For further information on Polity, visit our website: politybooks.com

Contents

Acknowledgements vi
Preface by Vinciane Despret vii

Preamble: Enforesting oneself 1

Chapter One: The signs of the wolf 13

Chapter Two: A single bear standing erect 34

Chapter Three: The patience of the panther 54

Chapter Four: The discreet art of tracking 93

Chapter Five: Earthworm cosmology 127

Chapter Six: The origin of investigation 140

Notes 172
Credits 182

Acknowledgements

Thanks to all the friends who have contributed from near or far to the expeditions which are at the origin of these texts, as well as reading and commenting on them. Thanks to Frédérique Aït-Touati and Marie Cazaban-Mazerolles for their alert reading and their generous feedback on the manuscript. Thanks to Stéphane Durand for his trust and his friendship. Thanks to Anne de Malleray for giving me the space and the freedom to experiment with these new forms of philosophical tracking narratives.

Thanks to Vinciane Despret for being Vinciane Despret.

Thanks to Estelle, finally, who shares a lot of these explorations with me, as well as sharing the adventure of writing them.

Preface

'Where are we going tomorrow?'

Where are you going tomorrow, or the day after, or maybe next week, once you've reached the last pages of this book? Perhaps you'll be one of those readers who will have the wonderful experience of being touched, contaminated, infected by the impulses that animate it. I could have written: 'by the adventure that impels it', but I'm a little wary of the epic exoticism or predictable storyline which the word 'adventure' can convey. I could probably more accurately describe what Baptiste Morizot is proposing by the evocative term 'initiation'. Being (or becoming) initiated involves the idea of getting to know something or, more precisely, getting to know the art which makes this knowledge possible; and this idea itself takes us back through the centuries to the experience of participating in the Mysteries as practised in ancient paganism.

Thus, this book proposes to initiate us into a very particular art, which could briefly be defined as *the art of doing geopolitics by tracking down the invisible*. Certainly, put like that, it might seem scary – and you might well wonder if it is really sensible to ask someone to write this preface who hesitates at the word 'adventure' but has no qualms about combining 'geopolitics' with 'the invisible'.

Preface

Forms of invisibility: 'You cannot exist without leaving traces'

However, nothing could be more concrete, closer to the soil and to life itself than Baptiste Morizot's project. It is, quite literally, the most down-to-earth proposition that you could imagine, a proposition which requires putting on a good pair of shoes and walking, but which mainly impels you to learn again how to stare at the ground, to look at the earth, to read the copses, the trodden grasses and the dark thickets, to scrutinize the mud where marks and pawprints leave their trace and the rocks where they don't, to inspect tree trunks with bits of hair sticking to them, to scrutinize paths where droppings are plentiful in one place but not another. For this is how those we call animals, and who are mostly invisible to us, manifest their presence. Deliberately sometimes, or even without paying attention. Tracking things down, in other words, means learning to detect visible traces of the invisible or, to put it another way, it means transforming the invisible into presences.

As Jean-Christophe Bailly has remarked how, for a large number of animals, their innate way of inhabiting their territory, their 'home', consists in concealing themselves from sight – 'for every animal, living means crossing through the visible while hiding within it.'[1] Many of us have experienced this: we can walk in the forest for hours on end and not sense the presence of animals, or even remain completely unaware of their existence. We can imagine that this world is uninhabited, believing ourselves alone. Yes – so long as we don't pay any attention to the signs. But if we change the way we walk through different spaces, pay due attention to them and learn the rules that govern the traces, then we are on the trail of the invisible, we become readers of signs. Each trace testifies to a presence, to the sense that 'someone has been here before' – someone we can now get to know, without necessarily encountering them.

Preface

Geopolitics: 'Tracking is the art of investigating the art of inhabiting practised by other living beings'

And yet an encounter does take place. But the term 'to encounter' here assumes a somewhat different meaning from the one that immediately comes to mind; it undergoes a shift and, as a verb, takes on an inchoative meaning,[2] like those verbal forms that indicate an action that has only just begun – grammarians say of these particular verbs that they indicate *the passage from nothing to something*. So the type of encounter that Morizot describes falls into the realm of beginnings: tracking always has to do with the time before an encounter, a time which, in principle, will continually be played over again (as the time before *is* the very time of encounter); and it only ever addresses what is already slipping away (the *something* of the grammarians could just as easily return to *nothing*).

What the practice of tracking also makes palpable is that to follow is to *walk with*. Walking becomes an act of mediation. Neither side by side, nor at the same time: in the footsteps of another who follows his own path and whose traces are so many signs that map his desires – including the desire to escape his tracker if he has become aware of the latter's presence. 'Walking with', without simultaneity and without reciprocity, relates to those experiences in which we allow ourselves to be instructed by another being, as when we let ourselves be guided, learn to feel and to think like another (who is, perhaps – like the wolf sensing that he is being followed – trying to think like whoever is following in his footsteps, as we shall see). We then abandon our own logic and learn another logic, we let ourselves be flooded by desires that are not ours. And above all, when we imagine and think on the basis of the signs left by the animal, where its intentions and habits lead it, so as not to lose its track – above all, not to lose track of it: for what we learn from the art of tracking is not to lose what we do not possess.

We can therefore 'encounter' in the sense of starting to know, without necessarily being in the same place at the same time – getting to know each other. 'Walking with', at a later time and at a certain distance, in order to be better instructed. Summoning the imagination in order to stay connected to a fragile reality. This is what the American philosopher Donna Haraway beautifully defined as 'intimacy without proximity.'[3]

To encounter an animal by means of intervening signs then means drawing up an inventory of habits which gradually shape a way of living, a way of being, a way of thinking, of desiring, of being affected.

The form of investigation proposed by Morizot points, first and foremost, to a profound change in our relations with non-human beings. More and more of us want to find a different way of living with animals, to dream of renewing old relationships, of catching up with them, as the saying goes. But how? What do we need to do? What should we learn? How can we live with other beings who are, for the most part, totally foreign to us? In this regard, Morizot has noted, with a touch of humour, that ever since the 1960s 'we have been seeking intelligent life in the universe, while it exists in prodigious forms on Earth, among us, before our very eyes, but discreet in its muteness.'[4] We send probes and even messages to the four corners of the universe, and we walk through the forest as noisy as a troop of baboons on the razzle, which can only confirm our strange conviction that we are alone in this world. It's time to come back down to earth.[5]

Preface

That's where this investigation comes in. As a geopolitical inquiry, it attempts to find answers to the question of how to live together with non-human beings, no longer as a rather abstract dream of returning to nature, but concretely and practically. Of course – and Morizot does not forget this – tracking reconnects us with the oldest practices of hunters. Nor does he neglect the ethology which is itself inspired by those practices, an ethology which he can now draw on for his project. These practices are arts of attention. However, unlike tracking, they do not involve knowing as a prelude to appropriating; and unlike ethology, they do not involve knowing for the sake of knowing, but 'knowing in order to live together in shared territories'. By tracking, we attempt to rekindle the possibility of forging social relationships with non-human beings.

'We can change metaphysics only by changing practices'

Tracking, therefore, is an art of seeing the invisible so as to frame an authentic geopolitics. As we have mentioned, there is nothing supernatural in these invisible things even if each discovery involves a certain magic, that of tracking 'which flushes signs'. There is nothing natural about them, either: there can be no serious geopolitics based on Nature. For the term 'Nature', even when used in such trivial circumstances as those which make us say 'we're going out for a walk to get a bit of nature', isn't innocent. As Morizot writes, with reference to Philippe Descola, this term is 'the marker of a civilization' (not a very likeable one, he adds) 'devoted to exploiting territories on a massive scale as if they were just inert matter.' And even if we decided to move away from this 'heritage' dimension and asserted our desire to protect nature, for example, we would not escape the tacit implications of this term – that there is, in front of us or around us, a passive nature, in short, an object of action – or, even worse, a leisure spot or place for spiritual renewal.

So Morizot's project asks us to dispense with a metaphysics that has caused definite and palpable damage and that we cannot hope to patch up with a few good intentions. The first thing that needs to be revised is the old idea that we humans are the only political animals. (Indeed, we should be concerned about the fact that when we declare ourselves to be animals, this is often a way of laying claim to a quality that simply confirms our exceptionalism.) But wolves are political animals too: they know all about rules, the boundaries of territories, ways of organizing themselves in space, codes of conduct and precedence. And the same applies to many social animals. Morizot takes up, and extends to other living beings – for example, to the worms in the worm composter, whose habits are similar to our own – the idea that what we need to relearn are truly social relationships with them. Tracking, as a geopolitical practice, then becomes the art of asking everyday questions. The answers to those questions will form habits, prepare alliances or anticipate possible conflicts, in an attempt to find a more civilized, more diplomatic solution: 'Who *inhabits* this place? And how do they live? How do they establish their territory in this world? At what points does their action impact on my life, and vice versa? What are our points of friction, our possible alliances and the rules of cohabitation to be invented in order to live in harmony?'

'A possible detour to get us back home'

I have just referred, as does Morizot, to the worm composter and its worms as a site for social exchange. A site that also requires a detailed knowledge of habits, attention, alliances and compromises. This example is important because it tells us that becoming a 'tracker',

Preface

'becoming a diplomat' with animals, actually involves a transformation in ways of thinking, of reading signs and of attuning (recognizing and creating harmony between) habits and intentions. Tracking *may* involve travelling great distances or through forests, but it doesn't always require it.

After all, as Morizot says, tracking is above all 'an art of finding our way back home'. Or rather, he implies, it is an art of *finding ourselves at home*: but this 'at home' is not the same as before, just as the 'self' which finally finds itself at home has itself become different.

Tracking means learning to rediscover a habitable and more hospitable world where feeling 'at home' no longer makes us stingy and jealous little proprietors (the 'masters and possessors of nature', as seemed so obvious to Descartes), but cohabitants marvelling at the quality of life in the presence of other beings.

Tracking means enriching our habits. It is a form of becoming, of self-metamorphosis: 'activating in oneself the powers of a different body', as the anthropologist Eduardo Viveiros de Castro writes. It means finding in ourselves the crow's leaping curiosity, the worm's way of being alive – perhaps even, like the worm, feeling ourselves breathing through our skins – the bear's desiring patience, or the panther's replete patience, or the very different patience of the wolf parents of a turbulent pup. It means gaining access, as Morizot says, 'to the prompts specific to another body'.

But 'all this,' he adds, 'is very difficult to formulate, we have to circle round it.'

In the wonderful book in which he recounts his long friendship with a bitch called Mélodie, the Japanese writer Akira Mizubayashi discusses the difficulties that his adopted language imposes on his way of describing the relationship between him and his animal companion. He writes:

Preface

> The French language, which I have embraced and made my own over a long apprenticeship, stems from the age of Descartes. It carries with it, in one sense, the trace of this fundamental break that means it becomes possible to classify non-human living beings as machines to be exploited. It is sad to note that the language of the time since Descartes somewhat obscures my sight when I contemplate the animal world, so abundant, so generous, so benevolent, described by Montaigne.[6]

We inherit, then, a language which in certain respects accentuates the tendency to de-animate the world around us – as evidenced by the simple fact (to take just one example as highlighted by Bruno Latour) that we only have at our disposal the grammatical categories of passivity and activity.

To narrate the activity of tracking, as Morizot does, to narrate the effects of this 'finding our way back home', involved learning to get rid of certain words, playing tricks with syntax so as to account for presences or, more precisely, effects of presence, so as to evoke affects that flood through the body (joy, desire, surprise, uncertainty, patience, fear sometimes), to use the writing of the investigation in order to touch on what goes *beyond* this writing, as Morizot himself was touched while writing. He had to twist the language of philosophy, to defamiliarize himself from it, to poetically force the grammar, sometimes forge terms or divert their meaning (what he has elsewhere called *a semantic wilding*),[7] because none of the terms we have inherited could express the event of the encounter or the grace of awaiting it. To create, in other words, a poetics of inhabiting, an experimental poetics, out in the open air, with plural bodies.

Beyond all that this book teaches us about what animals can do, as well as the humans who go out to encounter them, beyond the concrete and highly innovative political proposals for another way

Preface

of inhabiting the earth with others, Morizot invites us to explore not only the close confines of our world, but the very limits of our language. To express the event of life.

Where are you going tomorrow? Actually, from the very first words, you will already be on your way.

Vinciane Despret

Preamble

Enforesting oneself

'Where are we going tomorrow?'
'Into *nature*.'[1]

Among our group of friends, for a long time the answer was obvious, with no risks and no problems, unquestioned. And then the anthropologist Philippe Descola came along with his book *Beyond Nature and Culture*,[2] and taught us that the idea of nature was a strange belief of Westerners, a fetish of the very same civilization which has a problematic, conflictual and destructive relation with the living world they call 'nature'.

So we could no longer say to each other, when organizing our outings: 'Tomorrow, we're going into nature.' We were speechless, mute, unable to formulate the simplest things. The banal problem of formulating 'where are we going tomorrow?' with other people has become a philosophical stutter: What formula can we use to express another way of going outside? How can we name where we are going, on the days when we head off with friends, family, or alone, 'into nature'?

The word 'nature' is not innocent: it is the marker of a civilization devoted to exploiting territories on a massive scale as if they were just inert matter, and to sanctifying small spaces dedicated to recreation, sporting activities or spiritual replenishment – all more impoverished attitudes towards the living world than one would have liked. Naturalism, in Descola's view, is our conception of the world:

Preamble: Enforesting oneself

a Western cosmology which postulates that there are on the one side human beings living in a closed society, facing an objective nature made up of matter on the other side, a mere passive backdrop for human activities. This cosmology takes it for granted that nature 'exists'; it's everything that's out there, it's that place that we exploit or that we tramp through as hikers, but it's not where we *live*, that's for sure, because it only appears 'out there' in distinction with the human world *inside*.

With Descola, we realize that to speak of 'nature', to use the word, to activate the fetish, is already strangely a form of violence towards those living territories which are the basis of our subsistence, those thousands of forms of life which inhabit the Earth with us, and which we would like to treat as something other than just resources, pests, indifferent entities, or pretty specimens that we scrutinize with binoculars. It is quite telling that Descola refers to naturalism as the 'least likeable' cosmology.[3] It is exhausting, in the long run, for an individual as for a civilization, to live in the least likeable cosmology.

In his book *Histoire des coureurs de bois* (*The History of the Coureurs des Bois*), Gilles Havard writes that the Amerindian Algonquin people spontaneously maintain 'social relationships with the forest'.[4] It's a strange idea, one that might shock us, and yet this is the direction this book wants to take: it's a matter of following this lead. In a roundabout way, it is through accounts of philosophical tracking, accounts of practices involving the adoption of other dispositions towards the living world, that we will seek to advance towards this idea. Why not try to piece together a more likeable cosmology, through *practices*: by weaving together practices, sensibilities and ideas (because ideas alone do not change life so easily)?

But before setting this course on our compass, we first need to find another word for expressing 'where we are going tomorrow', and

Preamble: Enforesting oneself

where we are also going to live, for all those who want to move out of the cities.

For several years, among friends who shared the practices of 'nature', this question raised itself. To formulate our projects, we could no longer say: we are going 'into nature'. Words had to be found that would help us break with language habits, words that would burst from within the seams of our cosmology – the cosmology that turns donor environments into reserves of resources or places of healing, and which sets at a distance, out there, the living territories which are in fact beneath our feet, comprising our foundation.

The first idea we came up with to describe the project of expressing 'where we are going tomorrow' in different terms was: 'outside'. *Tomorrow we are going outside* – 'to eat and sleep with the earth', as Walt Whitman says.[5] It was a stopgap solution, but at least the old habit was gone, and dissatisfaction with the new formula prompted us to look for others.

Then, the formula that imposed itself on our group of friends, due to the oddity of our practices, was: 'into the bush'. *Tomorrow we're going into the bush.* Where, precisely, there are no marked trails. Where, when there *are* marked trails, they do not force us to change our route. Because we are going out tracking (we are Sunday trackers). As a result, we walk through the undergrowth, passing from wild boar paths to deer tracks: human trails do not interest us, except when they attract the geopolitical desire of carnivores to mark their territory (foxes, wolves, lynx or martens, etc.). Carnivores are fond of human paths, and these paths are used by many animals, because their markings, those pennants and coats of arms, are more visible there.

Preamble: Enforesting oneself

To track, in this sense, is to decipher and interpret traces and pawprints so as to reconstruct animal perspectives: to investigate this world of clues that reveal the habits of wildlife, its way of living among us, intertwined with others. Our eyes, accustomed to breathtaking perspectives, to open horizons, initially find it difficult to get used to the way the landscape slips by: from being in front of us, it has now moved beneath our feet. The ground is the new panorama, rich in signs – the place which now calls for our attention. Tracking, in this new sense, also means investigating the art of dwelling of other living beings, the society of plants, the cosmopolitan microfauna which comprise the life of the soil, and their relations with each other and with us: their conflicts and alliances with the human uses of territory. It means focusing attention not on entities, but on relationships.

Going into the bush is not the same as going into nature: it means focusing on the landscape not as the peak for our performance, or as a pictorial panorama for our eyes, but as the crest which attracts the passing of the wolf, the river where we will certainly find the tracks of the deer, the fir forest where we will find the claws of the lynx on a trunk, the blueberry field where we will find the bear, the rocky ledge where the white droppings of the eagle betray the presence of its nest . . .

Before even going out, we try to locate on maps and on the Internet the forest track by which the lynx can reach those two massifs to which it is drawn, the cliff where the peregrine falcons can nest, the mountain road which is shared by humans and wolves at different times of the day or night.

We no longer look for walks to go on, or signs of hiking trails that we come across by chance, surprised that they exist, no longer really understanding their signage. We are slowed down: we no longer gobble up the miles, we go round in circles to find the traces, it some-

Preamble: Enforesting oneself

times takes an hour to cover two hundred metres, as on the tracks of that moose in Ontario which was going round and round a river: an hour of tracking, losing and then finding its trail again, speculating on where its next traces would be, finding ourselves right back where we had started, next to the fir forest where, as an animal with night vision, it was probably taking its daytime nap, if we are to judge from its very fresh droppings. We are going 'into the bush' – and that's already another way of saying and doing things.

It's not, of course, a question of finding a new word to impose on everyone as a replacement for 'nature': we just wanted to piece together multiple and complementary alternatives, to find different ways of expressing and practising our most everyday relationships to living things.

The third phrase that suggests an alternative to 'getting a bit of nature' occurred to me one morning while reading a poem. It's the phrase 'to get a breath of fresh air.' *Tomorrow we'll get a breath of fresh air*. What fascinates me about this formulation is how the constraints of language poetically suggest something quite different from what you mean – how the phrase almost makes you hear the element most opposite to, and most complementary to, air, namely the 'earth' which the ear can almost hear hidden in the word 'breath'.

To 'get a breath of fresh air' is also to be back on earth, earthly, or 'terrestrial' as Bruno Latour puts it. The fresh air that we breathe and that surrounds us, by the ancient miracle of photosynthesis, is the product of the breathing forces of the meadows and forests that we walk through, and which are themselves the gift of the living soils that we tread upon: the breath of fresh air is the metabolic activity of the earth. The atmospheric environment is living in the literal sense:

it is the effect of living things and the environment that living beings maintain for themselves, and for us.

To get a breath of fresh air: the earth is disguised in the word 'breath', but still perceptible – and once you are aware of it, you can't ignore it. And the magic formula then invokes another world where there is no longer any separation between the celestial and the terrestrial, because the open air is the breath of the green earth. There's no more opposition between the ethereal and the material, no more sky above us to ascend to, for we are already in the sky, which is none other than the earth inasmuch as it is alive – that is, built by the metabolic activity of living things, creating conditions that make our life possible.[6] Getting a breath of fresh air is not about being in nature and far from civilization, because there is nature everywhere (apart from in shopping centres . . .). Nor does it mean being outside, but rather being everywhere at home on the living territories that are the basis of our subsistence and where each living thing inhabits the woven web of other living things.

To get a breath of fresh air, however, is a bit demanding: urban life as such, disconnected from the circuits that convey biomass to us, disconnected from the elements and other forms of life, makes it very difficult to access fresh air. In the heart of cities, this means tracking migrating birds or practising the geopolitics of permaculture vegetable gardens on a balcony. It means wondering where this tomato came from so that I can smell the sun and the portion of earth from which it was born, and see that earth with my own eyes. It means activating mutualist alliances with the worms of the worm composter to which we donate the leftovers from our kitchen and the shreds of our hair, so as to see and circulate solar energy in dynamic ecological processes rather than hiding them in lifeless rubbish bins. It's more difficult, but even in the city you can get a breath of fresh air. With a

Preamble: Enforesting oneself

little eco-sensitive vigilance, the living land reminds us of itself. It's fascinating to feel how much we are connected to spring, how much it rises up in us, reaching into the very heart of the big cities, something we can see from a thousand little invigorating signs.

Being in the fresh air means simultaneously being enlarged by the living space around us when we take up room within it, and with our feet in the soil, lying on it as on a fantastic animal which bears us, a gigantic animal come back to life, rich in signs, in subtle relationships, a donor environment whose generosity is finally recognized, far removed from the myths that tell us we need to tyrannize the earth if it is to nourish us.

Being in the fresh air means being in the living atmosphere produced by the respiration of plants, since what they reject is what makes us. It means recognizing that the breath of fresh air and the earth are one and the same fabric, immersive, alive, made by living things in which we are caught up, mutually vulnerable – and thus forced into more diplomatic relations?

Being in the fresh air is, at one and the same time, an invigorating opening and a way of finding our way back to the earth.

The last word, the one which finally summed it all up, is a word we stumbled upon by chance. It's a word from Old French that comes from the *coureurs des bois* of Quebec. It's the way they expressed the idea of going off for a breath of fresh air, after each return to town to do their business. They would say: 'Tomorrow I'm heading off, I'm going to *enforest myself* ('*je vais m'enforester*').

Enforesting oneself is a twofold movement, as the reflexive verb suggests: we go out into the forest and it moves into us. Enforesting oneself does not require a forest in the strict sense, but simply a different relationship to living territories: the twofold movement of

Preamble: Enforesting oneself

walking across them differently, connecting with them through other forms of attention and other practices; and allowing ourselves be colonized by them, allowing them to enfold us and move into us – just as the pioneer forest fronts of the Cévennes pines advance towards the villages, covering the old pastures which are no longer maintained by pastoralism.

It was tracking, in a philosophically enriched sense, that set us on the path to this process of 'enforesting ourselves', which shifted our way of looking and living – a tracking associated with other practices, such as picking wild plants, which require a very fine sensitivity to the ecological relationships that weave us together into living territories. This 'eco-sensitive' tracking inaugurates another relationship with the living world, which simultaneously becomes more adventurous and more welcoming: adventurous because so many things happen – everything is active, everything is a little richer in strangeness, every relationship even with the bottom of the garden deserves to be explored; and more hospitable because it is no longer silent and inert nature in an absurd cosmos, but living creatures like us, vectorized by recognizable but always enigmatic vital logics, a mystery which can never be completely fathomed by investigation.

There is a Zen aphorism which to my mind suggests something of the trail that we are following here, this trail to enforest ourselves. There's a monk standing in the pouring rain, his back turned to the door of the temple, gazing at the mountains. A young monk sticks his head through the door of the temple, bundled up in his robe, and says to the monk: 'Come back in, you'll catch your death!' The monk answers, after a pause: 'Come back in? I hadn't realized I was outside.'

In a sense, in the old days, we were often bored when we found ourselves 'outside', in inanimate landscapes, seeking physical exer-

Preamble: Enforesting oneself

tion and picturesque views. From now on everything is populated, calling out to us, and we must live together in a great shared geopolitics. Trying, as amateur trackers, to become diplomats towards forms of life that dwell among us, but in their own ways. We could undertake to become 'intermediaries' towards all these living beings. *'Truchement'* is a nice word from Old French – it can mean translators or intermediaries, and is sometimes used to describe strange characters: it was the name of the young French *coureurs des bois* that the explorer Samuel de Champlain, when he landed on the Algonquin territory that was to become Canada, allowed to winter among the Amerindian tribes so that they could learn the language and customs of the so-called 'savages' and become diplomats between nations, now wearing frock coats and sporting feathers in their long hair.

We would need to become *coureurs des bois* of the same order, but this time dealing with different 'savages': to enforest ourselves is, as it were, an attempt to winter 'over there', to see things from inside the point of view of wild animals, of the trees that communicate, the living soils that labour, the plants that are akin to the permaculture vegetable garden. To enforest ourselves means to see through their eyes and become aware of their habits and customs, their irreducible perspectives on the cosmos, to invent better relationships with them. It is truly a question of diplomacy, since it involves a variegated people whose languages and customs are poorly understood, a people that is not necessarily inclined to communicate, although the conditions are there simply because we share a common ancestry (we descend from the same ancestor). To 'enforest ourselves', we will need an acrobatics of the intelligence and the imagination, and an indefinite, delicate suspense, as we try to translate what those plants and animals do, what they communicate and how they live.

Preamble: Enforesting oneself

The anthropologist Claude Lévi-Strauss argues in a famous passage that our inability to communicate with the other species with which we share the earth is a tragic situation and a curse. When asked what a myth is, he replied:

> If you were to ask an American Indian, there would be a good chance that he would answer: a story of the time when men and animals were not yet distinct. This definition seems very profound to me. For, despite the clouds of ink projected by the Judeo-Christian tradition to mask it, no situation seems more tragic, more offensive to the heart and the spirit, than that of a humanity which coexists with other living species on a land which they enjoy in common, and with which humanity cannot communicate. We can understand why the myths refuse to view this defect of creation as original; they see in its appearance the inaugural event of the human condition and not of the latter's infirmity.[7]

However, this 'defect of creation' is just one point of view: communication is possible, even if it is difficult, always subject to creative misunderstanding, always surrounded by mystery. It has never ceased to be so, except for a civilization that has disfigured other living beings and turned them into machines, forms of matter governed by instincts or an absolute otherness governed by relations of force.

If, however, Lévi-Strauss's definition of myth is the correct one, then tracking appears, enigmatically, as one possible way among others of experiencing and accessing the time of myth itself.

This overlap between animal and human, this metamorphic experience between oneself and the other can be found everywhere in tracking. To understand the animal's trajectory, you have to put yourself in its place, see through its eyes. You have to find the key

Preamble: Enforesting oneself

points, the convergences between ways of being alive, by following the animal in its tracks. You have to find the (even more) vital problematics in yourself. In order to find the wolf, you have to probe the problematics you share with the wolf, inside of yourself: you have to try and move outside the merely human way of living so that you can coincide with something else. You have to move along a path, for example. Certain animal paths are places where the human and the animal are blurred, because it is not possible at first glance to decide who dug them. A path is often shared, drawn and carved out by several species, including humans, and it is with the same wary eyes, and for the same reasons, that they choose them. The deer's trails are open paths; those of the wild boar become difficult to follow when the shrub cover becomes denser, because they are low down; those of the chamois are often too vertical, as this creature lives like a bird in three dimensions, and the verticals are just as natural as the horizontals; those of the wolf are optimal routes to investigate.

Large animals form a community with similar reasons for moving about, and an analogous way of changing place; they embark on the same search for the cleared path, the best possible passage, the stream where they can quench their thirst or just revel in the joy of the living water, the sunlight in which they can warm their skins after the cold valley, the coign of vantage over the valley which allows them to orient themselves a little and see what's coming, the shade in which they can cool off at midday, the detour round the peak. A wolf's trail always takes the path of least resistance. And that's why a human being will naturally follow an animal path (if the animal has a certain corpulence), and that's why, in it and through this path, there is something like a momentary blurring of the distinction between man and animal which proves how close they are, in the vital, lived experience of their pacing along the same trail. They see it with the same

Preamble: Enforesting oneself

eyes; they are mammals that open up the path with the same aims and the same ways of thinking and deciding. Despite the differences, despite the unapproachable strangeness of other forms of life, there is at certain points something like a community of vital issues. This is what becomes evident in forest tracking when, for example, we discover a lost track because we have guessed that it was towards that babbling stream over there that the animal went when the temperature soared, or when we know in advance that, on this pass, the wolf, imbued by the sovereign desire to make its territory known to everyone, will have left a mark, which indeed we find right there. In passing, we unwittingly experience the time of myth: a time when human and non-human animals are no longer clearly distinguishable.

Like any good intermediary, it is to be hoped that a diplomat who has gone to enforest himself among other living things, even for a day or two, will comes back transformed, made serenely wild, far from the phantasmal wildness we attribute to Others. Whoever lets himself be enforested by them should, we hope, come back a little different from his werewolf trip, turned into a half-blood, straddling two worlds. Neither degraded nor purified, just *other*, and able to travel a little between worlds, and to make them communicate, so that he can work to bring about a common world.

The earth, that is sufficient,
I do not want the constellations any nearer,
I know they are very well where they are,
I know they suffice for those who belong to them.[8]

I

The signs of the wolf

This is the story of one night. It's 24 July, I think; a night with the flock. There are two of us. We've come along together to keep the wolf from the sheep and understand the critical moment, the nocturnal drama in which the wolf comes into conflict with human societies – when the wild animal attacks the domestic animal. It was while talking with the shepherd that the idea came to me: why not go and spend a few nights with the flock? I could show my presence, be of some use, and experience what triggers the controversy currently raging in France and elsewhere concerning the return of the wolf. We are following the flock of sheep on the Canjuers plateau. It is home to one or more packs of wolves, very deadly to sheep: the ecological, historical and pastoral conditions act as a catalyst for the attacks.

Canjuers is a military camp off limits to civilians, with bombardments and passing tanks. As we walk, we can hear shells exploding, in a nature empty of humans. In the distance, ghost villages. From this desert, the fauna are exploding back into vigorous life.

We stop under a ridge, facing the summit of Mièraure. It's 10 p.m. We pitch the tent in perfect silence, communicating in the signs of Native American language, using our lips just to smile. This caution is necessary. I stay outside until 1 a.m., in the company of a red moon, above the flock of about 1,200 hardy sheep, adapted to the harsh climate. The poor soil and the heat of the day require the sheep to graze all night. At that time, large herds, as mobile as flocks of starlings,

The signs of the wolf

are at the mercy of predators. There is sporadic barking from the guard dogs – seven *patous* mongrels (Pyrenean Mountain dogs) and Anatolian shepherd dogs – and then all grows quiet. I settle down in my sleeping bag.

I re-emerge at 3.30 a.m. because the guard dogs are howling continuously, on the defensive. They can sense that he's here. They take up positions around the flock. I walk down the hillside in silence, towards the animals, my unlit flashlight on my belt. I've chosen my path so as to be downwind of the flock. The lavender smells strong, the moon behind me is spewing out a beautiful clear light. The cry of the dogs arouses a certain fear mingled with adrenaline. For several minutes I pause a few hundred metres above the flock, my mouth open to better capture the sounds around, motionless and without turning on my flashlight.

Then I hear something, something coming up towards me. Zigzagging through the loose stones, crossing my path a few dozen paces ahead. I flinch, imagining a *patou* aroused by the conflict. I know those stories about guard dogs ripping the arm off a compassionate vet.

I first see him running along obliquely, sixty paces ahead; he's charcoal grey in the light of the moon, his coat is quite smooth; his shoulders roll powerfully in a way I have never seen in a dog; he's very long, his tail is low and very straight; he gives off a sense of power, the lucid power of a wild beast. Forty paces from me, he suddenly stops.

He's sensed my presence. He turns his head towards me.

He stares at me for two long seconds, the time it takes for me to pull out the flashlight from my belt. I shoot a splash of light into his face, but he turns away before the rays hit him. He darts up towards

The signs of the wolf

a thicket. I charge towards him, then change course to cut him off. I want, I think, to frighten him, to scare him away, to drive him away from the flock. But I'm not sure of the meaning of this highly instinctive act. Maybe I'm charging so as not to be afraid.

Has he picked up my scent? I was downwind, but the wind was a little unsteady, and weak. It was my silhouette in his peripheral vision that, I think, caught him by surprise. So he looked at me as an equal. Face to face.

He disappears behind an isolated grove on the edge of Gourdon Wood. I enter it. The thickets are dense. I pass under the foliage of black pines, which create a sanctuary, and I search for a few minutes. He's vanished.

Both my senses and my mind predict that he must be there: he's not.

I realize that my conduct is absurd and is actually creating an unnecessary danger; if he's lurking here, there's no point in cornering him.

So I come back and stand between the flock and the wood. Philippe, the shepherd, told me that the wolf often comes back to look for a slain animal. The wolf will wait for the flock to move away a bit, so that the dogs will scatter, or he will try his luck again. He doesn't reappear. He couldn't do so quietly, for the loose stones here prevent any silent movement. That's what surprised me most: the very loud noise he made while running. He's a wild beast, but not a feline.

I roll a cigarette while sitting in the lavender. I worry that I arrived too late. He seemed to be coming up from the flock. We'll have to wait until daylight to see the damage. I should have come earlier. It's really silly, considering the intensity of the attacks here, but I'll be happy if I've helped give the sheep a night's peace.

The dogs are finally calming down, they've separated out on the outskirts of the flock, and they'll bark sporadically every five minutes to indicate their positions, to keep awake, to give each other courage. They've done an admirable job. In the dark, I'm moved by their paradoxical loyalty as guards. Tonight they have faithfully served the human being who raised them to protect their ancient prey from their *ancestor* the wolf – and to gain in reward the meat of their new protégé (when the shepherds give them the remains of the dead beasts). These role reversals are enough to drive you crazy.

In Native American sign language, the sign for the wolf consists of a V formed by the middle and index fingers of the right hand, starting at the shoulder, and advancing diagonally towards the sky. The sign for 'dog' is the same V, fingers pointing down this time, but moving diagonally *backwards*, in the opposite direction.

I ponder the scene.

4 a.m., an encounter with a wolf forty paces away from me, *man to man*.

It's absurd, but this is the first and clearest expression that comes to mind when I try to put it into words. This impression becomes a riddle to be solved. It's not at all the macho face-to-face that this hackneyed phrase suggests – which is why I don't understand why it comes to me so spontaneously. These three words are an intuitive attempt to grasp something quite different, but what?

I don't see his face, because I'm too slow in drawing the flashlight from my belt. (Lesson 1: train yourself to draw very quickly. Lesson 2: it takes extreme rigour, asceticism and unpredictability, silence and stealth, to surprise him.) I cannot make out the white mask of his lips, nor really his pointed ears, that I can just about half-see.

But he's looking at me; no, he's looking at my face; no, he's looking

The signs of the wolf

me in the eyes. 'Suddenly you remember that you have a face.'[1] This memory plays a particular role in the consistent feeling of having *met him*. Eye-contact with the wolf: what a philosophical enigma. Why do some animals spontaneously look us in the eyes? If they thought that we were bodies moved by physical forces, falling stones, trees, or if they didn't think at all, their gaze would move over the entire surface of our body, without finding our gaze. Here, the fact that they are looking us in the eyes indicates that they know something: there is an intentionality hidden behind our eyes, as if there were *something* to see, as if we really had a soul whose presence was betrayed in these mirrors. I don't know how to say it. Eye-contact reveals what these animals understand about who we are. They ascribe an interiority to us – and we struggle so much to be just as polite to them in turn, as their gesture requires: there is only one interiority recognizing another, among the rocks, the forests, the clouds.

Adolf Portmann, a biologist of appearances, says something about the primacy of animal and human heads in these encounters. He isolates a correlation between the degree of cerebral complexification and the intensity of appearance: according to him, the more cerebral the animal, the more it engages in appearance. The appearance of the head in some animals is of this order: as a visible organ, it is highlighted as the governing part of the body by a whole series of adornments, contrasts, symmetries. He adds: in animals, 'The highest expression of individuals, i.e. the possibility of manifesting their internal condition, is of use in helping them to find one another.'[2]

This is a hypnotic meeting, because it occurs in another dimension, the dimension of the night, which is not meant for humans; in it, forms evaporate so we can no longer identify beings or master the space we move in. A realm of impressions, where the body is driven by hearing and smell, archaic senses that are once again dominant. At night, the clarity of biological knowledge disappears. Anatomy can only be effected in bright light. We need another language: we see 'wolf impressions', space-time complexes, unfinished silhouettes where the imagination supplements the gaps in our vision. It is so natural to see monsters – werewolves.

Strictly speaking, I have met a 'lupoid', more than a *Canis lupus*.

But I'm certain he's a wolf.

How can I know this so clearly? What secret deduction has taken place behind my eyes, in the great buzzing life of the mind, a life too vivid for me to see it? It takes me several minutes to reconstruct the flash of reasoning by unfolding its premises. After leaving, I heard the dogs howling, giddy with excitement, around the flock. They all barked and scolded to scare away the predator, to show each other their positions, to buck each other up. But the one running towards me over the loose stones is perfectly *silent*. He has the manners of a prowler. I hear the besieged, but I *see* the silent besieger circling around the flock.

This is why I know instantly, and unconsciously, that he is not a dog. And then there's his colour, for all of Philippe's dogs are whitish. And there's his tail, held diagonally straight down as he runs, which makes it possible to distinguish the wolf from even archaic dogs, which have curved tails. And then the stealthy attitude of a hunter seeking something. And then that silence, examining, assessing: the ancestor of the dog does not bark.

In Native American sign language, the sign for the wolf consists of a V formed by the middle and index fingers of the right hand, advancing towards the sky. The same sign, posted in front of the eyes, with

The signs of the wolf

index and middle finger pointed towards the ground, means 'to stalk', 'to *hunt*'.

Why this feeling of having seen him *man to man*, then? It's the odd feeling of a *human* face-to-face meeting that explains, I think, why this formula springs to my mind, and not the idea of some silly confrontation between males.

Why are some animals more readily cross-bred with humans than others?

Is it because wolf and human are super-predators, both occupying the ecological category of secondary consumers in the 'biotic community'? Or because both are social, hierarchical mammals, able to adapt to most environments, from the Golan Desert to the Arctic Circle, tireless explorers, curious to learn new hunting or fishing practices? I don't know.

The herdsman of the steppes recognizes the wolf's strategic intelligence during his collective hunt and compares it with his own stalking tactics when tracking ibex and bighorn sheep. These two great social predators share a family way of life, the way they collectively feed their young and teach them hunting techniques, as well as engaging in a dispersed existence which prevents incest: the Kyrgyz people consciously note these convergences in mode of existence. A nomadic herdsman interviewed by ethno-ethologist Nicolas Lescureux said:

> I talk about the wolf to my children. (. . .) I do this to make them attentive, or to compare him with humans . . . For example, human parents give everything they have to their children, everything there is, you prepare it for the children. Wolves are the same: when they eat in the field, they don't digest, they come back to their lair and they vomit . . . Yes, he's subtle (*kyran*), he's predatory; we're alike.[3]

The signs of the wolf

But I venture this hypothesis after my experience of spending nights with the flock. For the witness senses, in the marauding of the wolf which seeks to deceive the dogs in order to isolate a sheep, a powerful intentionality, intelligent, evaluating and tactical, one with the means to reach the end, determined and stubborn.

His stalking technique is intriguing: he seems to have understood that dogs cannot move away from the flock, otherwise they will leave gaps; he will therefore 'roam around the fort', like the Plains Indians, using guerrilla tactics, seeking less well-protected flanks, ready to beat a hasty retreat if the dogs arrive. This is the typical martial tactic of nomadic archers, the Huns of Attila and the hordes of Genghis Khan. It is said in Mongolian folklore that these people learned their war tactics from wolves: to be very mobile, to advance on the enemy, to retreat as soon as the static adversary is too powerful, to shift ground and launch a dazzling attack elsewhere. It's a state of siege in which he uses, with forms of intentionality and intelligence that may be strange but both are visible, tactics that he seems to have tried out in the field.

Infrared cameras have filmed a pack of wolves that send out a scout to lure the *patous* away from the flock, to make them run; another scout then does the same, while the rest of the gang circles the flock to attack. In military strategy, this is a case study which could be taught to cadets.

'Man to man': the scene looks like a confrontation between humans, because you have to rely on the cunning, determination and power of surprise of your lupine opponent.

The wolf therefore becomes a subject. In Western naturalistic ontology, the animal is traditionally the passive object, a piece of animated matter for the spectating subject, namely the human. Perhaps this is mostly because the animal is generally converted into food or

The signs of the wolf

into an instrument for human use. But when a being deceives our vigilance or our predictions, when we meet him on his own path, he becomes the subject of which *I* am the object, in a form of local metaphysical inversion.

In Native American sign language, the sign for the wolf consists of a V advancing towards the sky. The same sign, the exact same sign, means 'human', 'from the Pawnee tribe.'

It's so rare to see a wolf. What a statistical singularity that I should come across him and scare him away . . . But it must be said that he wasn't expecting to meet someone by this isolated flock, without light, at 4 a.m., camouflaged in the wind, his groin and armpits rubbed with lavender to mask his bouquet of scent. Tomorrow I'll go to sleep in the scrubland, on the opposite side, to take him by surprise, motionless, flashlight ready to be drawn (next day's note: saw nothing, heard nothing).

I settle down in the hollow of the rocks, between the flock and the wood where he disappeared, in plain view in the middle of open scrubland. I will stay here until dawn: if he has killed, my presence will prevent him from coming back for his prey; or else from coming back to hunt.

A night has never belonged to me so much. I lie there smoking among the lavender, watching over the flock, hat over half-closed eyes, above the bleating of animals aquiver in the light of a red full moon. And the barking of the dogs, guards and gatekeepers of the flock, a mobile castle, protean like a school of fish, a castle besieged by the wolf seeking a breach in their defences. I name it Turning-Round-the-Fortress after an illustrious Athapaskan Indian chief.

I unfold the map under the moon. All around on Canjuers flourish

The signs of the wolf

the various localities: Gratte-Loup, Saint-Loup, Le Cros du Loup, La Plaine du Loup, La Laouve (perhaps a derivative of *'éouvé'*, the oak thicket).

The toponymy alludes to the omnipresence of the wolf (*le loup*, feminine *la louve*) in the past history of this landscape. And wolves return to precisely those places that bear their name. So is this what some people would call 'home'? The first wolf we tracked in the snow in January 2012 around Lake Trécolpas was a massive alpha male. His digitigrade trot formed a calligraphy of terrifying signatures. Leaning over his pawprints, we looked up towards the undergrowth with a precise sense of being *with someone*. The return of the wolf to our forests strips us of our seigneurial, unquestioned certainty that these forests are our exclusive domain.

Long before dawn, I scribble these twisted lines on the lunar pages of a notebook.

An encounter that calls up another memory of an encounter. A walk at nightfall, at this precise hour when you can no longer distinguish the friendly silhouette from that of the enemy, treading softly on the moss to avoid cracking and vibrating noises. The Boréon forest is a flying forest. In the soft grey of the evening, suddenly, the top of a stone soars up and becomes a tit; then it is a stump whose bark rises and spirals upwards, spreading its finch wings. Leaves, branches take flight before you can realize that it's a wren. In fragments, in shreds, the forest flies up and recomposes itself. The drizzle flushes smells in the same way that a dog flushes sparrows. The rising odours of humus in the falling rain; the encounter of earth, water, air.

We're drawing near to a pack's territory, near a mountain torrent. Shoes off: a thousand needles of ice melt in the skin, and suddenly,

The signs of the wolf

in the song-chaos of the torrent, a rolling noise which is not that of a surge of water between two stones. I strain my ears, open my mouth to hear better, and again the sounds make their way to my ear through the mazy sounds of the torrent, with its many voices. Something in my stomach then tells me that this is not another voice of water, a voice of the torrent thundering all around – it's the polyphonic howling of a pack of wolves.

There's something prey-like, an immemorial past preserved somewhere within me, somewhere near the rhinencephalon: it's the past of a hunted beast, and this beast freezes inside when it hears this howl, as it is suddenly taken back 40,000 years to a forest of Pleistocene ferns, standing erect among the foliage, nose to the wind. A field mouse on a stone in the sunshine, overshadowed for a moment by the sharp silhouette of the kestrel; a trout soaring in the crystal-clear stream air, grabbed from the current by the bear's patient paw.

Your pupils contract, a frisson as long as a steel blade slides from vertebra to vertebra. Looking for the precise word, the exact word to describe what's happening, the secret upon which the mind stumbles and which it repeats is simply: if you're scared, you're right to be scared. *Ecce homo*: a prey, too.

I strain my ears again – the howls are gone. They nestled perfectly among the roaring rosettes of the torrent. It was a cry without any war in it, just a musical, melancholy appeal to the human ear. I don't know if I really heard it. I turn to my partner. She's behind me, her feet in the torrent. Bent like a bow, her gaze intensified, she stands motionless, staring in the same direction.

In Native American sign language, the gesture of striking one's chest, right hand open, palm facing the sky, fingers outstretched, is oddly polysemic. Awkwardly, we translate it as: 'by itself'. Associated with the sign 'gift', it means a free gift, one which does not expect any gift in return.

How can wolves live here? This is where, as a child, I used to go for a Sunday walk. It's a mountain for tourists, a changing museum where hard paths interconnect tableaux of grandiose landscapes, an open-air farm with friendly animals. The wolf hasn't been here for half a century, he was excluded from the nature-entertainment-suburban area developed for and by the Thirty Glorious Years of the postwar boom. But following Aldo Leopold's mantra, 'one need not doubt the unseen.'[4]

Even after his disappearance from our ecosystems, the wolf was visible in the grace of the deer, like the echo of a very distant past. The grace of the deer is a gift of the wolves. By exerting predatory pressure, wolves are the operators of natural selection and thus produce deer that are more agile, livelier, more alert, smarter, more powerful. This extremely acute vitality, this unprecedented near-perfection, woven into its own ecological conditions, and sensed in the casual movement of the deer you encounter by chance as it grazes or slips out from the edge of the forest into the sunlight, is precisely what we call the deer's 'grace'.

This is, perhaps, an invariant of animal encounters: when you come across a wild animal by chance in the forest, a doe which looks up towards you, you have the impression of a gift, a very particular gift, without any intention of giving, and without any possibility of appropriating it. This is what, in phenomenology, is called a pure gift: no one wanted to give, no one lost anything by giving, and the gift does not belong to you, it can be given to others. We feel an improbable gratitude rising within it. Just the desire to give thanks for the unforeseen and beautiful event which at this moment exists

The signs of the wolf

and gives itself to your eyes. It is the same feeling of free, inexplicable giving that takes place when one hears the pack weaving their collective song with their feet in the torrent. There's a line in a poem whose name and author I have forgotten: 'truly all is offered and impregnable, among the blue sky the green earth.'

I look up from the notebook: the cries of the dogs are getting louder again. They must have smelled something. We all stay together, one big animal, on the lookout. Long minutes pass, maybe an hour, a long second, boredom doesn't exist. We disappear, our ego streaming out slowly, in concentric circles, towards all that is perceived. The flock, then the edges of the forest, the ridges and finally the sky enter within, and we are no more than this spider web of living things, a web which throbs; we are no more than the span of our gaze. (There's something in a human being that can fly.)

The dogs are silent again, and the tension of the flock-landscape subsides, my mind returns to the encounter. I have seen a few wolves while hidden in the car, during tracking sessions. Here, on foot, on this dangerous terrain, without light, it is a meeting experienced as a man-to-man meeting, between equals. More than with other animals. Why this impression of a mirror? I keep coming back to this riddle.

According to Konrad Lorenz, the founder of ethology, the appearance of mobility in living beings is correlated with the appearance of intelligence.[5] It is because a creature can move that thought develops, seeking *where* to go, *how* to get there and *why* it should go there. Ends and means. Intermediaries to reach the end. The basis of any theory of deliberate action. But the wolf is a hypermobile animal, forever on the move, this is its ethological singularity: it spontaneously travels more than thirty kilometres a day, it roams over its territory, in search of prey, curious about everything, and also to assert its sovereignty over commensals and rival packs. A Kyrgyz song describes in vernacular terms the ethogram of the wolf, its mode of existence:

The place where he will stop –
The wolf thinks about it every day
And where he stops
The wolf always celebrates [. . .]
Running without stopping,
The wolf crosses plains, mountains
And as soon as the wind picks up
He takes in the scent and flees.

And this way of looking like he's going somewhere, following secret but real criteria, gives him a very singular aura of intelligence.

In Amerindian sign language, the gesture which means 'by oneself' also means 'free', or 'solitary'.

The roebuck and the deer just have to bow their heads to find their food, or to search here and there for the grass they wish to graze. The wolf has to move around a lot, in several modes (random marauding, the reading of olfactory and auditory signs guiding his quest, the stealthy approach, attack, then retreat). For the same amount of biomass ingested, he must deploy a wealth of intelligence in action. If we follow the fragile hypothesis which correlates resourceful intelligence with mobility; and if the wolf is characterized by hypermobility, both quantitative and qualitative; then, well you can draw the conclusion for yourself.

In the living realm, the fundamental emotional tone is split. At both ends of the spectrum, one has to live in fear or live in hunger. To be hungry or to be afraid is the dividing line between two different ways of being-in-the-world for animals, which probably correspond to places in the food chain, the food cycle. According to Lindeman's law, only ten per cent of the biomass moves from one level of the trophic pyramid to the next level. That is, one tenth of the plant biomass circulates, through grazing, to herbivores. And then only a tenth circulates from the biomass of herbivores to carnivores, through predation. This first explains the proportional mosaic of our living landscapes: there are many more autotrophic plants than herbivores, and many more herbivores than carnivores. Unlike herbivores, predators must capture living beings that do not want to die, and they often fail in this quest (for example, wolves are estimated to be victorious in only about one out of ten hunts). Being hungry, then, is the lot of those at the predatory apex: those who, once they reach adulthood, are at the top of the food chain and no longer fall prey to anyone.

Fear, on the other hand, is the fate of the prey, in this case the ungulates.

The omnipresent haze of satiety and fear can be guessed from the slow and erratic pace of the grazing, and in the extreme mobility of the ears of the roebuck, those nervy, reactive motor skills, that interminable watchfulness that comprises his life.

Hunger and quiet royalty can be guessed from the lives of feast and famine led by the wolf or the eagle, where you are quite prepared to suffer from hunger, but where you are not afraid, since no one threatens you. A form of sovereign casualness that is the basis of the other grace, that of predators in their way of moving. Silhouettes drawn with evolution's subtle brush.

Why man to man, finally?

We must meditate on his propensity to effecting an impossible disappearance.

The signs of the wolf

This wolf that I passed is entering a grove where, by sheer force of logic applied to clear and distinct matters he ought to be. He's not there. Philippe, the shepherd, tells us how a hunter crosses paths with the wolf on the Bel-Homme pass: the wolf runs over the plain, in the centre of which is a single puny juniper, behind which he disappears. The hunter looks down for a moment to pick up his phone. He stealthily approaches, walks around the juniper to catch the wolf by surprise, since he can't be anywhere else. There's nobody there. Doug Smith, the Yellowstone wolf expert, tells in his biography how he's in an aeroplane with his pilot, looking down on a wolf, and then he winks and, of course, impossibly, the wolf has vanished into thin air. Evanescent. Anecdotes of this kind abound.

My wolf has melted into a bouquet of night. He knows how to disappear just in the place where our brain predicts he is bound to be present. It is an aspect of ecological and ethological interaction which poses a beautiful riddle for thought: how does he do this? It's quite rare in animals. He has the conjuring art of misdirection, one which should be analysed. Misdirection is the art of magicians who can show you their left hand while everything is happening in their right hand. Our eyes, combined with our spontaneous mental modules, spontaneously deduce the position of things according to their trajectory, their speed, their volume. If the ball is in the magician's right hand, and he throws it into the left hand, then it ends up in the left hand. It's so instantaneous and automatic that we think it's the *real itself*, sheer evidence, hard facts. But this is a perception built by the brain, through the cognitive processing of information. This is the secret of conjuring. Getting the viewer's mind to make deductions that are invisible *to itself*, as if it were reality itself – but deductions that are mistaken as to what is actually going on. Making him believe that he *saw* what he *constructed*. And deceiving his automatic deductions by

The signs of the wolf

being aware of his cognitive biases. You did not see the ball go into the magician's left hand: you unconsciously deduced it, but your eye thinks it saw it.

Based on this model, how does the wolf manage to disappear? What grey area of vision and the human brain does it use to mislead our senses in their alliance with our spontaneous deductions? What optical illusions? What articulated cognitive biases? And how much intention and theory of mind does he really have? I don't know. We should certainly not fantasize about his skills or lend him too much intelligence, but the point of the problem lies elsewhere: it's what takes place on the ground, in intimate experience, when he disappears before your eyes, which remains the enigma – the moment of greatest emotional intensity.

One night, I was on a 'wolf patrol' on the Plan of Canjuers (this is the name that my partner and I give to the night patrols that we often carry out; they consist in driving through places where we know that wolves often pass, since their pawprints have been tracked there: mountain passes, roads, forest tracks which are wolf highways, passages where they habitually resort, the familiar paths along which they roam to survey their world). There was a lot of military activity that evening, big fires, passing tanks. I was just about to give up, when ten metres away from me, on the side of the road, just past the ghost village of Brovès, in a meadow punctuated by a single small oak tree, I saw the profile of a large canine, nosing the ground, sniffing a mouse hole.

I suddenly brake. He lifts his head, he looks at us with just an atom of curiosity, then off he trots, off he trots, quite nonchalantly. He's a wolf. But I'm not sure, I want to be sure. Following the same absurd impulse as that night in the flock, I run after him. He's ten metres in front of me in the meadow. There is only a thicket of three or four trees in front of us, aspen trees. I hear him making a noise in this grove. I get there a few seconds after him. I search the thicket with my torch. There's nobody there. I scan the bare meadow all around in front of me, in the axis of his flight, in a semicircle about fifty metres in radius: there's nobody there. It's totally impossible. I wander round for a minute or two, flabbergasted.

For anyone who has experienced this scene, the *effect* of magic is obvious: in reality, in sad, repetitive, mechanical reality, a breach has opened up and swallowed the wolf. A world of spirits into which he has vanished.

The spontaneous cognitive effect is clear: he can only be a supernatural being, if he can mock the laws of nature like this. There must be a background to the world, hidden dimensions, passages, invisible shortcuts. The world becomes strangely re-enchanted.

The wolf's cognitive magic is associated with the devil in Christianity, and with spirits in animism. This is probably the complex product of an adaptive story, in a predator whose cryptic skills, his art of hiding and disappearing, condition the success of his hunt and his escape from human persecutors.

But I think I've finally figured out what happened that night. After losing sight of him in and around the grove, I return to the car. I climb in quite dumbfounded, telling my partner about the mystery; she confirms that she saw the wolf's white lip mask. I restart the car and after a mile everything becomes clear. Here's how I tell myself this story: the wolf trotting ahead of me went off to make a noise in the grove, but then he did not continue on his way through the great open meadow: he retraced his steps. He made his way around the thicket, and instead of running across the meadow in front of us, as I had imagined, as I had deduced, as the evidence commanded, he took a few steps backwards, *behind me*. And probably he lay in the grass,

The signs of the wolf

not far from the path I had followed him down, he lay down just behind me while I looked for him *in front*. That's why I didn't hear him still running, although I could hear him trotting through the tall grass before. That's why I didn't find him in the great meadow where he should have been fleeing, in all human logic, beyond the grove. And when I got back to the car, I must have passed him very close to him: he probably remained motionless, crouching in the grass, not far from the field mouse, on the path we had both taken, where he had started. In my opinion, this is the only credible explanation, as everything else involved a devilish cunning that has no part in my cosmology. He lured me over to the thicket and then walked around me, and he went back to hunker down where I would never look for him.

Of course, no one will ever know what he thought, planned or wanted. Of course, this lupine conjuring trick is not magic in the 'supernatural' sense, but it does constitute magic in the *lived*, minimal, very simple sense of the word 'transcendence': 'it lies *beyond* me'. A transcendence of cunning, of wit: to deceive a human so thoroughly requires cognitive magic. For magic is often just technique, but a *hidden* technique.

Here is yet another way in which he crossbreeds with the human race — with an intelligence that is supposed to belong to humans. If the characteristic of man is intelligence, then the living being capable of deceiving my intelligence must, according to an insane syllogism, be . . . a little more human than me.

From man to man, then.

The night ends and I start to think of the others who will now play their part, the masters of the day this time, as they return to the flock. How are we to understand the conflicts between human *properties* and the *wild* predator? Seeing him marauding around the flock, I have a feeling that the predation of the wolf on the sheep cannot be intelligible in terms of Roman law where 'you truly possess only *what you have acquired legally*'. In the language of the breeders, the wolf steals the sheep, he *makes away* with them. In our imagination, he is often imagined as a stealthy criminal, a cunning burglar, an arrant knave. According to our Roman law of property, of course, he is indeed a thief. But this is a misunderstanding between cultures.

We need diplomats between men and wolves, literally: werediplomats, like werewolves, to decode the wolf's exotic customs. For when we see him hunting here, as if this were his home, probing the weaknesses in the defence with the full force of his powers, moving like a sovereign, living according to his own standards of behaviour, it is clear that he's not at fault, and that he's not without his rights.

His logic is analogous to *other practices* of possessing and taking. The Christian monks already said of the Scandinavians who came on their longships that they were looters, according to Christian law. But in Viking law, formulated in between the lines of the *Konungs skuggsjá* (*King's Mirror*) we find something like practical rules for traders who go to sea, for whom another code is set out, another norm, which could be formulated in this command: 'You truly possess only *what you can protect*'. Everything else actually belongs to whoever has the strength and cunning to take it. It is right in these Viking customs to take what is poorly protected: what you cannot protect doesn't belong to you.

So looting is not a crime against the law, but its serene expression. Here is the strange right of the wolf, which we should neither exalt nor ignore. It is Spinozist natural law, in which my law goes precisely to the limits of my power: what I can do, I have a right to do. It applies to those outside the pack, because, in it, prohibitions and

The signs of the wolf

rules of symbolic precedence govern access to collectively hunted food.

In Native American sign language, the gesture of striking the chest, palm facing the sky, means 'by oneself'. The same sign, the exact same sign, means something like 'wild'.

At dawn, the crows indicate no carcasses, and we find no bite marks on the animals.

2

A single bear standing erect

Northwest of Yellowstone National Park, there's a lake called Grizzly Lake. There was a huge blaze in this area in the great fire of 1988. The entire plateau stretching to the lake from the road between Norris and Mammoth Hot Springs is covered in a lunar chaos of blackened lodgepole pine trunks.

Walking here is full of the promise of encounters. Unaccustomed to the anatomy of animal feet in the Americas, I scrutinize the pawprints in bewilderment. In one clearing, there's a dark puddle stagnating in the middle of the path. Barely recognizable, at the bottom of the water, I can make out the pawprint of a bear. The claws are clearly visible, but the flurries of the wind on the water make the print almost unreal. In the damp earth, a colossal trace has left its imprint. This is the first time I've crossed the path of this species. None of us in my group are indifferent to it. We spend long minutes hunkered down in a circle over the track; from the outside it must look as if we're indulging in some strange form of meditation. The pawprint is pointing towards the lake, our destination. We make our way up the hillside to the summit of the plateau. After a few hours of walking and crossing three rivers, we climb into a dense coniferous forest. The Douglas firs that were scattered here and there on our ascent now form a dense canopy; between them, a trail of humus twists and turns. A few metres ahead, a long patch of snow hasn't melted yet, despite it being the end of spring. We still can't make anything out on it – but

A single bear standing erect

then a few signs, a few flashes of strangeness suddenly lift us into a state of vibrant attention, an attitude of intense openness that had seemingly been forgotten. 'We stay alert and alive in the vanished forests of the world.'[1]

It is then that the animal trail reveals itself, right there in front of our eyes, in the snow, several tens of metres long: the calligraphic passage of a bear on the path, its extraordinarily massive, incomparable plantigrade forelegs, the impression that another somewhat boorish mammal is strolling along majestically, an animal who is also something like an animistic god. Precise identification is difficult. An ancient trick helps to distinguish the grizzly's pawprint from that of the black bear. With a twig, you draw a straight line from the base of the big toe and through the top edge of the sole. If the little toe lies above, it's a grizzly. If it's underneath, it's a black bear. Here, in addition, the very long claws leave marks well ahead of the front paws: it's an adult grizzly. We're on his trail.

Another method for distinguishing between types of bear is sometimes mentioned in the bars of the West: the black bear climbs up the tree where you've taken refuge so he can eat you; the grizzly uproots the tree so he can eat you.

On the trail of the grizzly

The spacing of the feet indicates a slow walk, because the hind foot lands in the pawprint of the forefoot. The more the bear accelerates, the more the hind foot moves forward ahead of the forefoot. The wide spacing of the prints here indicates a particularly massive individual, perhaps a male, as he is alone: at this time of year, many females are mothers with cubs.

The trail leaves no choice. We have to move forward, we have to

A single bear standing erect

follow it: here we form a procession, behind the big bear, climbing up towards the ridge which overlooks the lake. I can't determine how old the traces are. The major cause of brown bear attacks on humans is not predation, but their reaction to being surprised. A very dense forest now borders the winding path. The density of the bush makes us invisible until the last moment, and we're moving forward upwind. We might surprise someone at every turn. We need to keep talking. In the shops of Yellowstone, hikers are sold bells that chime with every step. This is a paradox for those who engage in tracking – for those who try to learn silence. I watch my feet unconsciously practising the furtive Native American fox walk, designed to make no noise, while I sing an aubade at the top of my voice as I make my way along the path.

Following the slow gait of the bear for several hundred metres, and seeing, in its characteristic C-shaped traces, the places where it has stopped, the trunks it must have climbed, the bushes whose scent we sniff in turn, all establishes him as a guide and puts us in his place, so that we follow in his footsteps, see through his eyes, from the inside of his skull. This is not the first time that I've had the inner experience of following the trail of the same individual for a period of time, something that gradually enables the tracker to get into the tracked creature's head. This precise empathy probably has its roots in our very ancient, evolutionary-honed tracking skills – the hunter-gatherer lives of *Homo ergaster* and then *Homo sapiens* for perhaps two million years.

As we move towards the lake, storm clouds gather above us. Gradually, I pick up the trail of the grizzly. The traces are more and more discreet, the terrain more and more difficult to read. The thunder makes us get a move on, and our vigilance lessens. The early twentieth-century naturalist John Holzworth, a specialist in grizzly

A single bear standing erect

bears, argued that they are aware of their tracks. He cites examples of bears turning back, or circling their own tracks – so as to ambush hunters along their own trail. The Craighead brothers, who are field naturalists, argue that the grizzly bears of Yellowstone are prepared to wait for several days, with their muzzles pointed skyward, until a heavy snowstorm begins; only then do they enter their hibernation den, so that the cosmos may hide their traces behind them.

Suddenly, we come out on a meadow bordered by the emerald mantle of Douglas fir trees. Behind it stretches the mirror of the lake in which the clouds are piling up. This is the religious silence before the downpour. A crane strikes up its hieratic song. We move forward. At the edge of the opposite wood, an immaculately white coyote perched on a stump stares at me and then disappears with its strange grace, like a local spirit. A few drops of rain are starting to patter on the lake and the brim of my hat. With my binoculars I explore the edge of the wood where the coyote has disappeared, in front of me, then, in a circular motion, I turn around.

And behind us, almost on the path by which we emerged into the meadow, there he is. A brown, almost red-haired grizzly, his unmistakable identity revealed to the world by the frontal stoplight of his face and the muscular hump of his shoulders. I murmur the word 'grizzly'; it freezes us. He doesn't seem to be paying us any attention. Maybe he hasn't seen anything yet. Suddenly he makes a move. I see his two powerful arms grabbing a huge stump. He shakes it, his muscles bulging under his fur. He shreds it with disconcerting ease. We're squatting down. He is less than a hundred metres away. From where we are, he seems to cut off any possibility of retreat. He nonchalantly plays his role as a cosmic force, among storms and torrents, scattering pieces of wood the size of humans. And then he turns his head and stares at us. We speak to him in a low, calm voice.

A single bear standing erect

His hearing is so acute that he can recognize human voices at this distance, just as he can read the emotional state they betray. The voice should be deep so as not to be confused with that of a juvenile mammal, more easily viewed by the bear as prey. Deep, but not aggressive, so as not to be confused with that of a potential rival.

In life, a human being is sometimes less worthy of interest than a stump.

He merrily gets back to work. Only his ears, mobile and slightly inclined towards us when we speak, manifest his awareness of our presence. We slowly retrace our steps, maintaining the greatest distance between him and us. The storm breaks behind us as we descend the mountainside, and there is in our bodies a strange chemical state, bracing, joyful and dark – something like the whiplash of a rather pure fear.

Giving meaning to fear

The bear, and the grizzly in particular, is a special case in large mammals. He's one of those who trigger a deep natural fear – justifiably so. A grizzly is likely to attack humans if caught by surprise, or hungry, or protecting its young, as is the case in the spring when female grizzly bears are the most dangerous; or else if it is obsessed with the need to replenish its reserves for hibernation in the autumn, in the behavioural phases of binge eating. The grizzly cannot survive the winter without food and drink unless it has accumulated sufficient reserves from summer to autumn. Fat is the key to winter sleep. If he lacks fat as winter approaches, his feeding behaviour becomes a frenzied bulimia, lasting up to twenty hours a day, indiscriminately. Even ferocity is regulated and meaningful in living creatures, if we are willing to pay attention to the significance and rhythms of that ferocity.

A single bear standing erect

A few weeks after my return from Yellowstone, on a trail I had walked down alone, one of the park's emergency doctors, an experienced hiker, was attacked, killed and eaten by an old male bear. Tales of the Frontier, such as those of Jedediah Smith or Hugh Glass, abound in anecdotes of violent encounters, often fatal for humans.

Yet fear is a raw emotional datum, which the psyche needs to metabolize for the world to have meaning. In some cultures, human symbolic thought takes advantage of this asymmetry of power to turn an encounter with a bear into a test of male bravery. This omnipresent *topos* in Western culture is a way of coding and structuring the ethological emotions of this encounter in ritual form. In Scandinavian culture, for example, the duel with the bear consisted of a warrior caparisoned with leather irritating the animal until it rose up, then slipping into its arms and, surviving its fangs and claws, stabbing its heart, which was now exposed by this very embrace. A strange mechanism was sometimes part of this ritual: a dagger was fixed perpendicularly on a board hanging round the man's torso. It was sticking straight ahead so that, in the embrace, the bear would rush up and impale its own heart on the blade. Legends sometimes tell us that the adversaries rolled together down the ravine, and ended up tending to their wounds, a few steps apart, on the banks of a river.

It is probably this romantic, simple-minded motif of the encounter as a test of virile bravery that unconsciously guided my steps when, in the week following this first meeting, I found myself systematically hiking alone in areas where bears had been seen. I moved furtively and in silence, in search of an archaic initiatory ordeal.

A single bear standing erect

A dagger on the chest

One morning, just as dawn is breaking, I take the path that leads to the plateau above Lost Lake. I've caught a glimpse of a bear climbing between the trees. By roughly estimating his speed and trajectory, I try to keep a good distance. He surprises me by being closer than I had thought; and he hasn't seen me. I approach with open arms, reciting in a low voice the *Smokey the Bear Sutra*.[2] He's scratching his back on a log when he hears me. He rises up lithely on his hind legs, maybe thirty metres away, and scrutinizes me. We mirror each other for a few handfuls of seconds. He appears to be a young male, maybe four or five years old. I already sense how stupid I am being: how misplaced and irrelevant it is to have decoded this encounter by resorting to the common idea that it's a test of bravery. He trots away.

No doubt his good-natured silhouette, his youthful curiosity are partly the reason behind this defusing of any face-to-face standoff. Another way of imagining things replaces in my mind the idea of confrontation: the figure of the friendly bear, the clumsy bon vivant, embodied in children's stories where his large size and massive weight are no longer grounds for fear but occasions for showing an awkwardness that make you laugh. The figure of the friendly bear is the reverse image of the rival bear. One fantasy replaces the other, and hides the animal from us. What should we replace them with?

The sun is now above the horizon. I have rounded Lost Lake along its wooded side and almost got lost in the very dense bush. I could come face to face with anyone in this inextricable undergrowth, and the options would be limited. As I glimpse the exit of the forest along a thin path that runs along the lake, an adult black bear, a massive one, appears in my field of vision in the distance. He's heading towards me, along this path which is my only way out. I show myself, and advance with the greatest majesty a confused primate can muster. With both paws raised, we begin a complex diplomatic ceremony, which I don't really understand – but it's as accomplished as if my life depended on it. It results in a mutual treaty of non-aggression. The bear snorts and changes its path, leaving the way clear for me. I finally come out on the meadow in front of the lake.

There's no question here of any test of bravery. It's about something else, another form of encounter – but of what kind?

The two options provided by Western culture for interpreting the encounter with the bear (as confronting a beast or greeting a friend) are each based on a biased conception of our relationships with living beings: on the one hand, the despotic myth which stipulates that we must overcome nature to civilize it; on the other, an Arcadian ecology that dreams of a nature without hostility. But wild animals are not our friends, as in the contemporary fantasy which establishes our pets as models of all animality; nor are they beasts to be conquered in order for us to accomplish our civilizing destiny. We need to look for other models, other ways of thinking about our relationships with them, such as their otherness.

A small path set deep in a valley leads from Lost Lake to the Petrified Tree layby where I left my car. It's now eight in the morning, and I don't know it yet, but there are three more bears on my path. A massive grizzly bear and, further on, two black bears. There's no other way of getting back without running the risk of coming nose to nose with the previous two. I may need to elbow my way out of the valley of Lost Lake. Engaging in dialogue, attracting the attention of the red-haired grizzly without looking him in the eye, negotiating, tapping my staff with the blade of my knife, thanking the big taciturn female bear who elegantly slips away, giving just a bit of a scare to the female baby bear lounging at the end of the path, I finally find myself all wrung out, tied up in knots, my senses incandescent, in the metal sanctuary of my Ford.

Only the blindness proper to the male could interpret these encounters as a test of bravery and empty the animal of its substance, until it is nothing more than a mirror in which virility seeks its own gaze – a mirror in which to measure its own strength.

No, they were not glorious rivals to compete with. They weren't amiable stuffed animals. They were superpowerful creatures out for a stroll hoping, like all other living things, for a good day.

That parable of the Scandinavian warrior with his dagger tied to his chest to test his bravery . . . what a strange way to imagine an animal encounter! Wandering down the paths, ready for anything, on the alert, ambling along at random, but with an *erect* dagger sticking out from your chest. A dagger pointing at everything that comes your way.

There must be other ways of testing your courage. For example, going out to meet other living things with so little fear that your violent aggression, which is only the mask of fear, dissolves, and gives way to diplomatic intelligence.

The diplomatic courage of these explorers who advanced palms open towards the foreigner, their weapons idle at their belts, but fully vigilant and on the lookout, capable of defusing the crisis by an extraordinary act of empathetic decentring, even though fear makes everyone self-obsessed – locked into his own point of view. The decentring which allows us to sense the ethology of other creatures, and, with the delicate force of intelligence, to impose a peaceful outcome on a confrontation which always risks turning into conflict.

A single bear standing erect

Perhaps this is another, albeit ancient, way of presenting yourself to living things.

The relationships that some First Peoples have with the wild animals they encounter on a daily basis, in all their beauty, their strangeness and their diversity, can be a guide here.[3] For animism or shamanism, these cohabitants of the Earth demand a singular form of respect: they are not our enemies, but they are not our friends either. They exclude any familiarity in interaction, and spontaneously demand a certain modesty, and something like an informal ceremonial – as towards a proud and foreign people who share this world with us, and whose enigmatic proximity elevates our conception of our own existences.

The politeness of the wild

Diplomatic courage towards brown bears would involve mastering an etiquette of the wild – one fully stocked with a knowledge of foreign customs, the rules of ethological politeness, the protocols of a fine intelligence (how to behave when fishing, hiking, or on horseback, what to do with the food in your bivouac, when faced with a bear or a bear cub, when faced with a carcass, how to dispose of your waste when living in a bear area . . .). You have to have these skills at your fingertips, and then adjust them to the singularity of each encounter, like a good diplomat.

The customary ceremonial requires that you modulate your body language so as not to appear like an aggressor or a victim – a delicate balance. It advises you not to seek eye contact, but to look at the bear obliquely, and never run because this is a stimulus that, in the bear's ethogram, triggers pursuit. Some of the advice given by the experts is almost comical: 'If the bear charges at you with all its might, don't run away. The charge *could* be a bluff.' There is indeed a kind of diplomatic courage. Standing firm in the face of intimidation when your adversary's jaws are wide open may be the most appropriate response.

Pepper spray, finally, is a truly wonderful diplomatic device. Worn on the belt, to be unsheathed in an instant, it looks like mosquito spray intended to ward off any aggressive quarter-ton colossus. But intelligence is extremely effective, ethologically speaking: the bear weighs perhaps five times as much as us, but its sense of smell is a thousand times greater than ours. In other words, so is its sensitivity. However, capsaicin, found in chilli peppers, irritates the mucous membranes in proportion to their sensitivity. A cloud of gas can, at the press of a thumb, normally stop a grizzly bear in the middle of its charge.

It is still the function of diplomacy to prevent conflict from taking place, to maintain the conditions for mutual cohabitation. The last resort, if all negotiations have failed, is still of course to defend oneself physically, with any object capable of causing harm, and 'by all means necessary', as the instructions say (and this is also, not by coincidence, one of the watchwords of the United States Marine Corps).[4]

But then again, empathetic intelligence is the best shield: some experts advise that, if the bear brings you down, you play dead. Others say you defend yourself as fiercely as you can. What are we to make of this contradiction? These two opposing attitudes probably make sense, but each according to *one* ethological context that needs to be deciphered: when a grizzly charges at you, this amounts to a claim for dominance or a fight for territory; he is indeed more than likely to spare a mammal that presents him with the ethological signs of immobile submission. But when a binge-eating grizzly attacks you, this is for the purpose of predation, and he isn't going to be inhibited:

A single bear standing erect

a prey that plays dead will simply make his task easier. The whole problem is to keep on the lookout with that empathetic finesse that can make all the difference, and there are probably few diplomats who are capable of it.

To stop the encounter turning bad, some bear experts suggest that you talk to the bear politely; after all, you would expect a stranger who blithely turns up in the middle of your living room to explain his intrusion. In 1976, while patrolling Glacier National Park, Agent Clyde Fauley accidentally came across a grizzly female with her two cubs fifty metres away. The three bears charged and stopped about ten metres from him, growling, ears laid flat on the napes of their necks, their jaws clacking, and all three rose up on their hind legs. Fauley decided not to run to the big tree ten paces away from him. Instead, he called out to the bears in a low, calm voice, saying, 'OK, everything's fine here bears, this is no big deal – we guys in the park are on your side.' Meanwhile he slowly moved backwards up the trail towards a patrol cabin, and began to recite to them, point by point, the park's bear management policy. On the official report of the incident, he wrote: 'My conversation with these bears probably would have sounded ridiculous to an outside observer. Nonetheless, I believe it saved the day.'[5] No warlike confrontation, for this would end in the eternal victory of man over animal, but rather ethological negotiations no longer *dictated* by fear.

For this is an ethological law of the human animal: the bellicose posture is dictated by fear, which takes hold and asserts itself only when a human feels overwhelmed by the situation: when he gives in to fear. A less reactive courage, like Fauley's, does not consist in denying fear (which is an ineluctable given), but in denying fear the right to determine your behaviour: it consists in letting fear roar within you – without roaring yourself, without taking fear as the truth of the situation, without it eroding the attentive smile and the egoless, decentred, alert intelligence by which you open yourself to the possible peaceful outcomes of any confrontation. Paradoxically, this is often a more effective way of coming out of it unscathed; above all, it's more effective for living with *others*, whoever they may be. It's the strange courage of confronting otherness without concluding that, because otherness is dangerous, it constitutes an absolute enemy. It's a form of empathetic courage that fathoms the other's point of view, sees from all eyes together, sees from the point of view of the relationship itself. It is a courage without exaggerated virility, because it is a non-gendered courage (and it's actually an ecofeminist philosopher who provides the model that we shall be discussing shortly). A perspectivist courage which consists in confronting the other without bestializing it.

The lesson of fear

We can indeed decide to try diplomacy, but fear does not go away. It must contain a lesson. The emotional intensity of these encounters must be metabolized for the world to be meaningful. If the bear is not disfigured into a mirror of bravery, or as a beast to be defeated if we are to civilize the Earth, but instead recognized as a cohabitant in a shared biotic community, demanding etiquette and diplomacy, what lesson can we draw from fear?

My guess is that the lesson involves the bear's status as a *man-eater*. If the latter is no longer a challenge to males, he takes on another dimension. The lesson of fear does not say, 'Go and prove your courage.' But it could instead mean: 'Remember that you are also prey.' That is to say, you irreducibly belong to the food chain, to the ecological systems that create you and keep you alive.

A single bear standing erect

'Man-eater': there's something immemorially emotional about this phrase. David Quammen maintains that the relationship of fascination and repulsion that we maintain with regard to man-eating animals lies in the way they remind us of a part of our human condition that we have forgotten, or rather obscured thanks to our control of predators: the fact that we are also meat (this is the animal part of the rational animal).[6]

But what does this reminder of our vulnerability mean, how can we understand it in all its implications? On this question of being meat, in 1985 a crocodile enabled us to make great philosophical advances. He had the presence of mind to attack, on a river in Kakadu National Park in Australia, not just any hiker, but a philosopher: Val Plumwood. Alone in a canoe, far from any human presence, she was paddling down the river. A crocodile then struck her boat multiple times. She tried to escape by leaping towards a tree leaning out from the shore. The great saurian grabbed her in its mouth as she jumped, seizing her by the crotch, and subjected her to the underwater vortex by which it suffocates, stuns and annihilates all of its prey's energy and desire to escape. She managed to stay clear-headed when she emerged from the whirlpool and, released by the monstrous mouth, again grabbed at the branches to pull herself up. As she came out of the water, she was caught once more in the saurian's mouth. Three times over, she was immersed in the vortex. She kept her cool, and sensed, like some zoocephalic diplomat, that she needed to change her tactics and stop fleeing upwards, because her movements seemed to constitute a stimulus prompting the crocodile to attack. It was as if terror had not dented her intelligence and, with the empathetic courage of continuing to think in the midst of chaos, of decentring herself where fear and suffering usually lock people up in their own selves, she took the decision that required the greatest strength of soul: she stopped struggling – and allowed herself to slip away, motionless, in the current. The trick worked. She then managed to reach a muddy bank and painfully haul herself up it. It would take her several hours of wandering, seriously injured, but, guided by her knowledge of the bush and her survival skills, she finally managed to get help from the park rangers.

This tragic experience provides us with valuable testimony: that of a philosopher whom a great predator kindly reminded her of her condition as a biomass that can be shared by others. She is an authentic philosopher, that is to say an expert in perplexity, a professional in the inversion of perspective: she has the eye of a falling Icarus, but one almost capable of pulling herself out of the equation and considering the metaphysical stakes of being devoured *herself*. She quickly demonstrates her power of decentring: she does not let her fear react with blind aggression. In my view, Val Plumwood embodies diplomatic courage. A few hours after the attack, injured and weakened, she was taken to the hospital in Darwin by the rangers; here she overheard them saying they wanted to go and kill the monster, and it didn't matter which monster: killing any of them would do the trick. This was Val Plumwood's reaction: 'I said I was strongly opposed to this idea: I was the intruder, and no relevant purpose would be served by random revenge. The river there was full of crocodiles.'[7]

What she sensed was that the rangers were not out to incapacitate a dangerous animal (something both reasonable and necessary), but to react with fear, to reconstruct a certain order of the world, and to avenge the violation of a *taboo*. This event gave her a front row seat from which to witness one of the founding taboos of the Western conception of nature:

> It seems to me that the Western culture of human supremacy is characterized by a very great effort to deny that we humans are also animals

placed in the food chain. This denial that we are food for others is visible in our mortuary and funeral practices. The solid coffin, which is buried, as convention dictates, well below the level of soil fauna activity, and the slab above the grave to prevent anyone from digging us up, prevent the western human body becoming food for other species.[8]

What this critical experience produced for Plumwood was the shaking of the foundational myth of a certain Western humanity in its relationship to nature – a humanity that had defined itself as the product of a self-extraction from biotic communities, thereby inventing an 'outside', namely Nature.

However, this phenomenon of extraction is impossible in practice: we are living beings caught in the food chain, and like all others we must devour the sun.[9] We can't extract ourselves from food chains without starving. Being unable to feed ourselves directly from solar energy, we must wait for it to be captured and converted into biomass in plants, and subsequently in the herbivores that graze on them, before we can draw our daily vital energy from it.

Piecing together a myth

So how can a founding myth be credible if it is fantastical? How can we believe that we have extracted ourselves, freed ourselves from nature, when we will always be attached to it by the food chain? The hypothesis that interests me here is that in order to turn this *impossible* self-extraction from food webs into a self-fulfilling myth, at some point in history, human beings in the West had to develop a cosmology and a taboo that postulated a *diodic* relationship to the food web, which can be expressed as follows: we are allowed to feed on the sun trapped in living creatures, but other living things do *not* have the right to feed on the sun trapped in us. Diodic: this adjective defines the diode as a device for circulating energy (which we are, in fact, in the metabolic and ecological senses), one which allows energy to pass in *just one direction* – here, from the world to us, and not from us to the rest of the living world.

The taboo therefore consists in prohibiting or minimizing all the conditions in which we would be biomass available *to others*. The drive to eliminate superpredators in a certain section of Western culture, which has often been ascribed to the need to protect livestock farming and the identification of predators with the devil in the Judeo-Christian tradition, here assumes another meaning: their destruction is a mechanism that maintains the taboo. They will no longer be able to feed on our remains, or our living bodies. The taboo is necessary to make this myth of self-extraction believable: it is necessary for real experience to coincide with the fiction, for the fiction to become true without being contradicted by the facts. The events in which the human is then 'reduced' to the status of meat constitute a foundational transgression, which calls for reparation, to restore the order of the world.

However, in other ontologies, being eaten does not trigger the same psychoses. In the cosmology of Siberian shamanism described by anthropologist Roberte Hamayon, the world order is described as a circulation of flesh. When he feels his death coming, the elderly person goes into the forest where death takes him. He thus gives back his flesh – his remains are shared by carnivores, so that it circulates, in cycles of reciprocity, back to the forest which provided him with it through the countless hunted beasts of prey which served him as food.[10] Other cultures also consider being eaten as part of the order of things and not a cosmic transgression. This can be seen in the sky burials of Tibet, where the mortal remains are made available

A single bear standing erect

to wild vultures and canines, as the deceased's self-gift to the land that gave birth to him. The terror of being consumed is therefore not *universal*: this is the clue that allows us to see in it the taboo linked to a foundational myth.

Western humanity, by contrast, has invented itself as a diode for cosmic energy: the only species in which the circulation of energy, or of the flesh-sun in the living cosmos, is solely a *one-way process*.

Val Plumwood writes: 'This conception of human identity places humans outside and above the food chain, not as guests at the feast in a chain of reciprocity, but as external manipulators and masters of this chain: we can eat animals, but they cannot eat us' (my translation).

We can be eaters, but not eaten. Uneatable eaters. The trophic pyramid is not a pyramid by chance: the geometric pattern was originally used to represent Lindeman's law. As a tenth of the biomass is captured from one level of the food chain and taken to another, this implies that the members of each higher level are much less numerous than those of the lower level, which can be symbolized by this triangular shape. But within the framework of the myth, the trophic pyramid is hijacked and used as a reason for transcendence: it clearly shows how to achieve this univocal relationship with the rest of the community of the living. Indeed, in a trophic pyramid, each level is caught up in symmetrical relations with the one above and the one below, the eater of the lower level being eaten by the upper level – each level, that is, *except the top*. Here, relations only go in one direction. Occupying the top of the pyramid is the transcendence of those without transcendence: monopolizing it is the only way to make ourselves believe that we have extricated ourselves from a biotic community of which we are actually members.

This positioning at the top is then methodically constructed by the destruction of large predators, as well as by multiple mechanisms for preventing the circulation of the biomass of the human body (burial six feet underground, tombstones, rot-proof coffins, etc.), and by tales that spread horror-stricken panic at the thought of being eaten. Our definition of the human condition can then exclude the fact of being meat, thus metaphysically ensuring our extraction above the biotic community.

In our place

It is this myth of origins that Val Plumwood debunked in her tragic experience. This demystification becomes the lesson of fear. Learning to live with large carnivores such as bears or wolves takes on a whole new dimension: 'Large predators test our ability to accept our ecological identity. When allowed to live in freedom, these creatures are a sign of our ability to coexist with the Others on Earth, and to represent us in reciprocal and ecological terms, as members of the biotic community.'[11]

This ability to coexist is not wishful thinking or a matter of natural harmony: it does not mean that we will let ourselves be eaten without defending ourselves. It requires all our intelligence to devise shared habitats, and to exercise a diplomatic behaviour capable of minimizing all risks to humans – minimizing them without having to resort to the widespread eradication of others with the alleged aim of pacifying the Earth. Big predators are mostly territorial animals. Territoriality was invented by evolution as a mechanism for pacifying the relations between living beings, a mechanism that is much older than human laws and conventions. The unbridled ferocity of these animals is a myth of the moderns: they can be ferocious just as they too can seek to decrease conflicts and aggression, especially when we ourselves, equipped with our particular intelligence as animal diplomats, can

A single bear standing erect

establish the conditions which allow for cohabitation, and make room for other creatures.

The forests of Yellowstone appear in a different light. On my last morning, at length, travelling through his eyes, I track a large bear around Grizzly Lake. I don't have to see him.

A single invisible bear transforms an entire mountain range, imbuing it with another gleam. He gives relief to each bush, which now has a hidden side. He digs a new depth into the thickets, which regain their dimension as *habitats*. He stops Nature becoming the backdrop for a selfie. He brings out other poles in the world, because we are no longer the only subjects, the only point of view that configures the world: a touch of fear, even if the risk is very low, forces us to recognize that there is another subject which objectivizes us, simply because he can *treat us as an object*, that is to say, subject us to his will against our own desires. He gives us back our ecological status as one living creature among others, caught up in the great circulation of solar energy that constitutes the biotic community. He reminds us of our diplomatic obligations towards this community, on which our lives are based. Nature again becomes that plurality of points of view that it was until we eradicated all the great predators and set ourselves up as a single point of view onto an inanimate nature that we have flattened out as a matter without mind, a nature reduced to the status of resources that are ours to dispose of, a nature that we have failed to see as itself, treating it instead as a mere mirror for ourselves.

A single bear standing erect can enable the whole living world to arise behind it.

3

The patience of the panther

Day 1

This morning, out on the steppe, sitting astride our Kyrgyz horses, we are investigating the fauna and flora of the Song Kul Lake Nature Reserve in central Kyrgyzstan. We gallop along, each tied to his own mount, sharing the joy of running with the horse; we gallop to each ridge, watching the sky behind us rise up, every time new, always the same. Just for a moment we rediscover the old habit of animals which disperse and, with each generation, settle further away (those animals capable of becoming accustomed to many environments, such as wolves, ravens and humans). The habit of always exploring new territories, just wanting them to be somewhere else. Travelling: coinciding with an immemorial gesture of living creatures, a gesture which we have, like many others, appropriated as human – one which involves going to see what lies *behind*. What lies beyond the limits of our gaze. This is an infinite task, because the limits move with our gaze.

If we think about this in the light of the data of the history of our species, it is because we are, like wolves and crows, animals that disperse; we, like them, have colonized the entire Earth, from the deserts to the Arctic Circle – and not because some singularity of human beings, an awareness of and thirst for the absolute and elsewhere, has made us curious explorers driven by an abstract greed for discovery.

The patience of the panther

It is because we are animals which disperse that we have a taste for travel, an unquenchable thirst for elsewhere.

After a whole day on horseback in the steppe, however, where our birdlike gaze exhausts the prospects as far as the horizon without being able to land anywhere, the yurt finally offers a round enclosure shut off from the outside, where our attention can curl up like a raptor in its nest. The yurt of Osmon Baiké, a ranger of Song Kul Lake and our overnight host, is like the leather hood that Kyrgyz falconers use to cover their eagles' heads and extinguish the vibrant attention to the world that their super-powerful sight forces them to maintain.

Tonight, a snowstorm pounds the protective layers of felt made from the carded wool of the sheep bleating just outside. There are no trees in the steppe: the only fuel for the fire that makes human life possible is condensed sheep manure, dried and cut into bricks. To live in the fur of animals, heated by the fire that only they can provide, invigorated by their flesh, and to sweep across the plain astride their calm and powerful bodies – this is certainly already another way of relating to them.

Day 2

The next day, at the entrance to the Naryn reserve, in southern Kyrgyzstan, we divide our packs among the horses that are to take us out on an expedition. More than eleven days in total autonomy, on a protected reserve, a sanctuary for all the living creatures that inhabit the ridges, the high steppe plateaus and the spruce forests. Only rangers and scientists have the right to enter it.

Going to see what lies behind. What lies beyond the limits of docile experiences. The objective of the expedition is to monitor the wildlife of the reserve, still poorly understood, by engaging in participatory science practices. It is based on a form of eco-volunteering. We will apply the techniques of scientific ecology: observations, readings of indices of presence, counts and transects, hikes along very precise trajectories connecting GPS points and repeated identically at different periods of the year to note any potential changes. Large predators and raptors are going to be a major focus: we will be locating the nesting areas of the golden eagle, counting the Himalayan vultures, tracking the bears, monitoring the wolves and, above all, relentlessly searching for the creature that gave its name to the NGO which has organized this expedition, OSI Panthera (for Objectif Sciences International), the snow leopard, the 'ghost of the mountains'. We will need to find the snow leopard's traces, her immaterial paths, from the floors of the glacial valleys to the snow-capped peaks. Seven amateur explorers of the living realm, with the support of four of the reserve's rangers, in this deserted, trackless place, where not all species have yet been found (the two ornithologists among us will discover three hitherto undocumented species of birds).

We will go up the River Naryn, which flows far behind us into the River Syr-Darya, and draws its sources from far ahead of us to the west, in the Tian Shan, the heavenly mountains, called in the Turkish language Tengri Tagh: the mountains of the sky god.

The expedition will meander along the river, along animal paths far from any human settlement, and climb up among the spruce forests which shelter the Himalayan brown bear and the red deer, to the high plateaus and the snow-capped peaks over 4,000 metres high which are the privileged habitat of panthers and ibex. The wolves, of course, are at home everywhere.

In the evening, in the last shepherd's house before the wild forest, we eat the piquant stew of mutton and rice, our mouths burned with chai, clumsily trying to communicate with the rangers who are accompanying us, the hosts in whose service we have placed ourselves, who speak only Kyrgyz and Russian. Our guide and translator, Bastien, is the most perfect of interpreters, but there are more and more stammerings and stutterings and silences when faced with the unformulable. Wrapped in their military parkas, the rangers laugh affectionately, the way they do at everything that happens – life being, of course, a great opportunity for a laugh while gathered round a cooking pot.

Djoldosh Baïké, affectionately known as Djoki, is the vice-president of the reserve. He is accompanying us as a bear specialist and expedition leader. At nightfall, in the military tent, after having swapped funny stories populated by Kyrgyz drunkards in exchange for a few French jokes, he falls silent for a moment and asks our guide to translate a question for the 'philosopher': 'Was ... was nature made for us, or are we, with her ...' (he searches for a word in Kyrgyz, which he ends up finding in Russian) 'partners?' He continues, stirring his tea: 'I think we're partners. But we crush her under our boots like an anthill.'

Day 3

In the morning, we load the horses again and dive into the reserve; the abundant rains have caused generous grass to explode everywhere and all the flowers draw the bees to them.

We form a caravan and follow the river, which rolls along in a powerful, almost clay-grey stream under the blue sky. On the path, in front of the horse, there is a flash of strangeness: a massive black dropping, the first riddle. It takes us a few tens of seconds to decide that it is indeed bear excrement, but not typical. It is studded with spruce thorns. Djoldosh, with his elegant camouflage outfit, his

The patience of the panther

golden-toothed smile, his childhood dream of being a wild animal tamer and his degree in bear biology, tells us that the Himalayan brown bear *does not eat* spruce in summer.

On closer examination, the strangeness increases: chitin – the material from which insects' shells are made – is strewn over the dropping. The bear is an omnivore and as such it shares with us humans, with the red fox and the jay, the animal destiny of being a tireless *taster*. What had he been exploring this time?

There then begins a long session of tracking on horseback to solve the puzzle. The bear likes this path: the droppings accumulate every hundred metres. Each dropping has this same strange content. Our desire to understand sharpens with our gaze, which now searches everywhere for signs. 'The world is all clues. There is no end to the subtlety and delicacy of the clues. The signs that reveal are always there. One has only to learn the art of reading them.'[1] Leaning over our horses' necks, keeping them on a short rein, we track clues from our saddles, animals on animals tracking every kind of animal. Djorgo, my horse, doesn't understand the situation: I'm sending him mixed signals, placing my weight forward looking for tracks, a gesture which tells him to 'speed up', and keeping the reins short, which means 'slow down'. Out of the corner of his black eye, he sees my face leaning over his neck, scanning the ground in front of his hooves. And then his attention tightens, as if he can feel that I'm looking, his curiosity is pricked and he searches with me, ears cocked forward, nostrils dilated, perplexed: I can see him asking himself 'what on earth is he looking for?'

When you track from above, with a bird's eye, the paths and trajectories are drawn out on the ground, the clues are configured like symbols on a map spread out in the grass.

In the sand, a few steps away, a new flash of strangeness electrifies

The patience of the panther

me: a dark, structured stain. This is the print of a bear, a forepaw with the mark of just the paw's ball. A small *Ursus arctos isabellinus* (three to four years old according to Djoki). He's moving in the same direction as us, and we follow him. Suddenly one element of the landscape becomes significant. For a few kilometres, we've been moving past anthills. But one of them, just there, has a strange shape, like a decapitated pyramid. The collective hypothesis emerges: this, then, is the meaning of his happy journey – he's touring the anthills. He decapitates them to feed on the tiny Hymenoptera, and in the process ingests the spruce thorns which *they* use in large quantities to form their admirable shelter. These wood ants often leave very visible flows, or highways, leading right up to the anthill: this is a blessing for a tracking bear.

In my notebook I write: 'Tracking comes down, for him as for us, to following an animal logic.' For that, you have to know him a bit. For example, we sometimes look for bear tracks in small steep valleys, because, in springtime, the bear examines the avalanche flows while licking himself – in case an ibex, swept away during the winter and preserved by the cold, is thawing in the sun.

The trees where the bears scratch their backs are recognizable: they draw our eyes, as they're often clearly placed on a path, an alluring prospect, with hairs clinging to them in abundance. When a bear strips trees, he makes horizontal tooth marks, while deer marks are vertical.

Once we have formed the hypothesis of the anthill, our entire field of vision is reconfigured: we come across more than fifteen disembowelled anthills. We're on his familiar path. The droppings are fresh. The trace on the dry sand was recent, not erased by the river or its edges eroded by the rain. He's out there somewhere. The wind is on our side: it drives our scent behind the rumps of our horses. Something joyful and dark tenses itself in the possibility of an encounter. In silence, on the lookout from our horses which now grow attentive, we dissolve completely into the surface of our gaze, which is like a bird in front of us exploring each valley, each mountainside. We pass from peak to peak, our gaze each time sharpened by hope. The traces accumulate. On a small plateau, the wind turns for the first time and delivers our scent to a forested slope above us. Our scent has been picked up. I scrutinize the scene: the sudden arrival of so many mixed smells could make him come out. 'Ayou!' shouts a ranger, pointing at the top of the hill (*ayou* is the Kyrgyz name for the bear): a golden flash of fur has sprung from the forest: he gallops along, almost flying on the green grass, in the late afternoon light. He crosses at length, moving away from the source of the scent, as quick as a mountain goat. He's a youngster. Maybe three or four years old. We all fill our gaze with this vision, in silence. We hit the road with our hearts lifted, laughing, humming – what a strange power he possesses.

A line from the Persian poet Omar Khayyam, who composed his quatrains west of the Silk Road on which Naryn was a stopover, comes to mind: 'Life passes, that mysterious caravan – you must steal its moment of joy.'[2]

Day 4

When we wake up the next morning, there is hoar frost on the tents, and a clear sky. Tea with milk from the thermos at six o'clock. We leave immediately. We head off in search of raptor nesting areas. The eagle feeds at dawn but waits for the ten o'clock thermal currents to climb high in the sky: so, at that time, he flies very close to his breeding ground. The common kestrel, on the other hand, utters a

The patience of the panther

characteristic cry, a cry of excitement, near the nest, revealing his secret.

At noon, I go alone with Djoki, at a full trot, to pick up the camera trap he left last year next to a real trap, a large metal tube baited with meat to capture a bear so that a GPS collar can be fitted onto it for research purposes. The trap has not yet been activated; it is set up and filled with fragrant meat several months before the scientific mission, so that bears will get into the habit of feeding on it without fear.

We slip the memory card into a camera: dozens of images scroll past. Among them appear, punctuated by Djoki's joyful exclamations, snapshots of two bears, a female and a large male, then a weasel, and another mustelid. The she-bear has been photographed several times: the sequence of images reveals the meaning of her action: she walks around the trap, she goes behind it, she explores its mechanisms in perplexity. She's looking behind the scenes.

We join the caravan at a gallop on the cliff-top paths, a few centimetres from the ravines that fall steeply into the waters of the Naryn; then a large meadow welcomes us at the end of a riverbend. It is covered with umbellifers larger than a human being; their white flowers form a continuous carpet reaching up to the rider's waist. Only the horse's head emerges on the surface as it gallops through this lake, spreading scents and fragments of flowers amid the cavalcade of hooves.

We slow down. All around, we survey the ridges – our eyes are stalking the bear, the wolf, the huge red deer. Then, leaning forward, with our hats for faces, our gaze tracks the slightest clue, trace or pawprint, in search of puzzles to solve.

The sky has turned cloudy by the time we catch up with the caravan, then the clouds burst. We ride through the forest under the hailstorm, the soil covered with white pebbles of ice, calligraphed in black by the horses' hooves. In my damp notebook I write: 'The rider's poncho protects the horse from muzzle to tail, and its rising heat envelops me under the cloak – in an exchange of good deeds.' The convoy stops: the barely visible path has been swept away in front of us by an overflow from the river. The members of the team ahead become active, breaking open a path in front of the horses with shovels and pickaxes, sawing the trunks, cutting down obstacles with an axe: opening up a way, fighting against the forest's obstinate habit of endeavouring to close itself. Constructing the path against the inexhaustible erosion of the world.

Day 5

The days pass. We arrive at the Umeut hut (pronounced 'Ü-mert'), a log cabin whose roof is covered with meadow grass. This will be the base camp for a strenuous expedition: we will climb up to the Umeut ridge, at an altitude of 3,900 metres, and follow it in a transect several kilometres long to detect all possible traces, especially those of the panther, which is thought to inhabit this area.

We load the horses with the tents, and head off to sleep just under the ridge, in landscapes defying the imagination, endless bars of rock straddling the sky. The climb begins, over 1,000 metres of vertical drop, through myricaria bushes and scree. We stop frequently to scan the slopes, the Kyrgyz rangers detecting with their naked gaze the invisible ibex that we can barely make out with binoculars.

Their observational finesse is such that, for certain species, they do not use a generic term: they have a word for the females who, outside mating periods, live in united herds, and another for the males whose horns are more curved and massive. Their ibex (*Capra siberica*), is here called *etchky* (for females) and *tekey* for males: *etchkytekey* in general, but this word is rarely used. The French language also once

The patience of the panther

possessed this subtle art of seeing, and expressing what is seen, but it is now forgotten: the French word for 'ibex', in fact, is *bouquetin*, the contraction of *bouc* (the male goat) and *étagne* (the female). The memory is there in the language, we just need to bring it back up into our bodies and our gaze.

Evening is falling and the sky is clouding over; we are a hundred metres from the small plateau lined with rich grass which will host our camp.

My horse is an ambler (*djorgo* in Kyrgyz): he has had this elegant and comfortable gait since birth. He's a little black horse and I've nicknamed him Kara Djorgo, after a mythical horse in Kyrgyz culture, 'the black ambler', the equivalent of our Bucephalus. My steed isn't quite such a handsome beast, with his ultra-skinny rump and his waspish waist, but he has the courage of the myth. The Kirghiz say that amblers are complicated horses: most roam every corner, exploring, venturing here and there, always wandering up hill and down dale; you have to hobble them first when you arrive at camp in the evening, in case they drag the other more docile horses off with them on their travels. In addition, our mounts, which had been placid until then, are starting to feel the vigour of the mountain itself swelling within them, and becoming animated by its stubborn vitality. As Takou explained to me, at high altitudes, there are more ultraviolet rays, so the grass is richer: for a few days, the horses have started to gain weight, they are revivified, they want to run around more spontaneously as soon as the meadows open up in front of the rider. In the morning, they are harder to catch. The sun communicates with the steppes of the high plateaus, and changes the behaviour of horses, in those loops of interdependence which form the basis of the subtle life of ecosystems.

The patience of the panther

Horses here are more than a means of locomotion, and something other than friends. Yesterday, I wanted to join the rangers who had taken the lead in rebuilding a section of the trail that had been washed away by a flood. They had taken my horse. I found myself wedged between two rivers, which I had crossed the day before on Djorgo's back, without even thinking about it, taking advantage of his confident step, of the mobile throne – that intelligent balcony – which he offers to the rider. Now, without him, it was impossible to cross. The feeling of helplessness turned to gratitude. The horses here free us from the limits of our bodies, the bodies of bipedal primates, adjusted to flat, stable ground: they make of us another animal, a multiplied animal, prancing like the winds. They possess, and grant to us, the surefootedness of the ibex in the rocks, the breath of the wolf, the panther's ability to withstand cold, the cunning of the bear.

These vital powers arise from the fact that the Kyrgyz horse, although domesticated, is a rustic animal, which is farmed on site under conditions chosen by the mountain itself: it is said that the mortality rate of domestic horses here is similar to that of wild horses. That is, the Kyrgyz horse is drawn, in each of its features, by the ruthless and delicate brush of natural selection, which here bears the beautiful name of 'winter, wolves, and free pastures.'

We climb again, under the black sky, towards the ridge of Umeut. Djorgo is tired, he's panting, the slope is too steep. He has Gandhi's skinny buttocks – and his courage.

I write in my notebook: 'Kyrgyz horses, convicts slaving on behalf of science'; and that's when the storm hits us. It's also the moment when Djorgo decides that he's not going any further. Time to put on my poncho, as the caravan disappears on the ridge above. I dismount,

The patience of the panther

crushed by the waterspouts, and endeavour – cajoling and begging my horse like a capricious god – to pull him behind me to the top. I almost have to hoist him up to the plateau, where everyone is busy setting up camp under the icy water. Undeniably, we put on our best tent-pitching performance. And then the storm stops, the sky clears, and we're probably in the centre of the world, since all around us takes our breath away, and elevates the strange sky inside of us that some call 'soul'.

Every day here works on us like the wind works on the silhouette of trees, by their sensory intensity: the icy gusts that cut across the whole body, as if we were ghosts, as if we too were winds, but we continue to move forward, transfixed, then invigorated now that we have become intangible; the hailstorms that nail us to our horses, and the joy of moving forward: all those spells of bad weather that we had learned to flee, which here for a moment frighten us, but which we get through (as there is nowhere we can escape to or find shelter in), until we finally understand – transformed, wrung out, blown through, soaked to the skin – that there was nothing to fear.

Then the scorching suns dry us up, and everywhere, all around us, the immense landscapes enter our eyes as if with forceps, making us as vast within as the land itself; and the icy rain again, the lurching of the crevassed soils and the bush, the horse's amble along the giddy ledges, the crossing of the rocky ground, the torrents, the thickets, countless soils for the feet to adjust to, exercising the most intelligent parts of the living body.

It's a physical intensity too: walking over lands unacquainted with any paths, kneecaps wobbling as we stumble over scree, on crossings not made for humans, then washing in icy water, our skin battered by stinging junipers, but always joyful, as we are raised above ourselves, always quiet, after having unlearned a little more each day our unfounded fear of hail, rockslides and giddiness.

From this submersion we emerge tautened by a new sensory intensity, our inner space swelling like a ship's sail when the wind rises, a sail unfolded on the scale of the sky god of these parts.

The next morning, we find our first panther pawprint, perfectly drawn in the clay, on the ridge line, which we follow between thunderstorms and suns. Then, finally, we descend on foot from this lonely ridge, empty, a windbeaten patch of flat colour, an inaccessible kingdom to which we will never return.

(And what is there on Umeut Ridge besides wind and memories?)

Day 7

The next day, we split into two groups: those least drained by yesterday's expedition are invited to set off on horseback, with two rangers, including Djoki, to explore. They head to a high cave, far up in the mountains, where the elders say that the bear dwells. There are two volunteers, my twin sister and myself. I let Djorgo rest in the meadow along the river, and strap my things onto another horse, an elegant brown bay, a very endearing creature.

It's the story of a typical day for the Kyrgyz ranger. Rising with the dawn, we prepare the horses, and begin the endless climb, in a one-eyed valley perpendicular to the Naryn River. First the meadows rich in flowers, towards the foothills, and in the distance the mountain corries, crystal clear, and the cave we were told about the night before. For hours, we make our way up on horseback. Djoki is tracking a bear that has come down a stream; he takes samples from its droppings, and notes the GPS coordinates of any clues. And we set off again, peripatetic scientists, towards the summits. The horses

The patience of the panther

climb up through vertiginous scree, towards the cave whose dark mouth we can now see above us. Our mounts paw at the ground, retreat, slide down. In the middle of a mass of fallen rocks, in a place where nothing can be seen, on an impassable slope, Djoki turns in his saddle, and smiles at us: 'Eat', he says. It is three o'clock. We dismount on the spot; our horses are immobilized by the steep slope, they doze in the sun. Leaning against the rockfall, we take out bread, cheese, sausage, chocolate. All around us, fantastic creatures made of mountains larger than entire cities eat from the sky's hands.

We stare our eyes out peering into the telescope, still tracking along the ridges the silhouette of the ibex, the movement of the panther.

As a light wind picks up, we set off on foot towards the cave: even the horses could not go any further. There is no path, of course, to allow us to approach, no human here has left a mark. You have to scramble up. The cold wind has this old habit of rendering inhospitable a wall which just a minute before was quite benevolent. The dark cave entrance is right here; we hoist each other up. The cave is cool, shallow, lifeless. And then, as our eyes get used to the dim light, a universe of signs appears. On the ground, ibex droppings. Here on this ridge, the rock is polished, it is barely visible, our fingers confirm: is it a panther which has rubbed its cheeks and its side, depositing a mark whose meaning is for us enigmatic, intended for its fellow panthers and other passing travellers? Djoki breathes in deeply the scents of the wall, almost adhering to its surface, his paws resting against the rock. He turns around, his forehead creased.

That's when we find the droppings of a large carnivore. Crouching around it in a strange ritual, at the back of a cave, as far from everything as if we were in another world, we examine it in silence. '*Irbirs*', says Djoki eventually. Panther. So we set up a camera trap, at

The patience of the panther

the bottom of the cave, placed to photograph all in its view, up to the entrance that overlooks the sky and nothing else. We take samples, record the GPS coordinates, and off we go, in an exhausting mixture of deserts and mountains.

It's four o'clock. The horses have regained their courage. They take us through the rockfall too steep for human feet, to the pass between Tchongtalde and Kichitalde. There's nothing here but stone and snow. The pass is so narrow that the horses are mere silhouettes as they make their way through. We explore the ridge, whipped by a very pure wind, stepping on the blade of a knife sharpened by the sky itself.

I am scanning the surrounding ridges through binoculars, when, electrified, my attention takes on an almost inhuman intensity: I've seen something moving whose 'jizz', as birders refer to an animal's singular appearance, resembles that of a panther. It's disappeared behind a boulder. It's a mystery. Who was it? The shadow of my desire to see?

Here there's nothing, says the human being. A desolate ridge, nothing but rocks and snow, a desert of solitude. But a few signs catch the eye, a certain art of looking, arousing an intuitive sense that the wilderness for us is home to a multitude of other creatures. As the gaze becomes sharper, entire habits, sovereign and orderly, emerge: on the ridge itself, the barely visible trails of the panther, revealed by a few hairs deposited under the boulders at the crossroads of paths, territorial markings. A hunter on the lookout, it moves silently above its prey, the ibex, the large Asian ibex. Then, a few score metres below, the ibex's own path appears, revealed by the droppings but still barely visible amid the scree. We counted the ibex in the two valleys on either side of the pass. In the first, the one we arrived

The patience of the panther

through, they were scattered (4, 2, 2, 1, 1). In the second, scanned with a telescope from the pass, they're in compact groups (47, 27, 32). So there are wolves in the first valley: it is in their visible reflection in their prey's behaviour that we can best distinguish the stealthy predators: the ibex adjust to their presence by dispersing in small groups to limit the risks. The scope of the invisible is immense, and yet it's in their reaction to what is invisible to us that we understand what is visible to others, what drives them and prompts them to act.

At the bottom of the pass there now clearly appear the wolves' markings: their excrements, their coats of arms and flags. Ibex inhabit inaccessible rocky ledges, under the panther's ridges, where they are most secure; for below, where the waves of the steppe come to break amid the scree, wolves reign, marking their territory with all these droppings. Ibex have their own modus vivendi, performing a balancing act between the panthers of the ridges and the wolves of the valleys, and their defence consists in surrounding themselves with a barricade of rock, as they are the virtuosos of verticals. The landscape is structured so precisely, the signs are everywhere: the habitats are intertwined, everything is populated, the desert does not exist, there are only shared homes. So discreet that we have forgotten them.

Animal habits, in fact, rarely transform the landscape dramatically, in the way our roads and our homes do. They leave clues and tiny, essential traces. Our blindness to the fact that our living spaces are also the habitat of others comes down to the fact that our human habitats are characterized by this material structure that transforms matter. From this, we deduce that the others do not inhabit a place like us, not really: birds do not transform the sky, nor dolphins the pathless sea, we think. But this is the bias of a technically-minded primate: other animals also inhabit, just in a less obvious way. Their way of living emerges through tracking, which reveals their familiar paths, the paths which organize their home range in a very refined way, connecting water points, places for nesting and brooding, dormitories, points of view, areas for play and display . . . Habits are the animal way of inhabiting space, of appropriating it, of making it familiar, turning it into a home (as is still so clearly visible in human animals). They are the animal art of arranging and developing their territory. The body of habits is the invisible habitat of the non-human animal: its intangible arrangement of the territory. Visible paths are just the traces of this process. As soon as you become aware of them, it becomes possible to make hypotheses about key points in the living world – and to set camera traps there.

So we set up two camera traps in the middle of this absent crowd and Ardak, the youngest ranger, climbs on all fours down the barely visible path of a fawn on the ridge, with all the grace of the panther, to ensure that his path does indeed pass down the axis of our machine for making works of art. I reflect that the photos taken mechanically by the trap in our absence resemble those works that the history of art calls '*acheiropoietes*': that is, 'not made by human hands' (and of course, usually, this word tends to be used to describe the works produced by divine miracle, such as the Turin Shroud).

The art of setting a camera trap, even on a ridge in the sky, lies in finding an inconspicuous point on the most familiar path of a living creature. If the choice is perfect, the deserts will almost certainly send you images of their inhabitants: just imagine hiding a camera trap between your mailbox and your doorstep.

On the pass, 4,000 metres above sea level, each sensation has a biting purity to it, a weightlessness that is dizzying. We walk in single file along the tightropes that are the familiar paths of the panthers, those knife-edge ridges which are the home sweet home of the big beasts, their imperial panoramic view of their prey and then of the whole cosmos – and there, under my feet, I catch the gleam of a rusty metal object. It's the case of a poacher's bullet.

There's a euphoria of the high mountains, a biology of spirituality: you feel accelerated, as if caught up into an intensified pace of life. Your red blood cells, the rarefied oxygen, your incandescent senses, the disproportionate verticals, the tests you have to face – all restore to bare existence a rusticity stripped of everything accidental, in the ragged wind of parasitic thoughts, and even confer strange sensations of purity.

Looks like it's time to come back down.

Holding our horses by the bridle, we launch out on foot into the scree on the other side of the pass, towards the next valley. We prance around, laughing with fear in the scree, dragging our horses along behind us until they get caught up in the game. I can feel the joyful breath of my horse's nostrils on my neck, and its hooves dancing behind me.

But always our gaze rises towards the ridges, towards the summits. What is the reason for this magnetization? As if moved by a life of their own, our eyes seek out the panther, the Himalayan vulture, the wolf or the ibex. They long to see these beings who live at home in this rocky chaos, in all the majesty of their inaccessible temples, those homes of the living gods.

At the bottom of the scree, a névé blocks our way: a large patch of snow straddling a torrent. The atmosphere tightens imperceptibly on Djoki's cheekbones. Ardak sets off to reconnoitre, leading his horse in slow motion, each step of the hooves on the snow as delicate as a stitch. Suddenly the horse plunges sharply into the snowfield, right up to its chest. The torrent roars just under the snow, and fifty metres away is an endless waterfall. I can't help but turn to my sister, and an unstoppable smile grows on our burnt faces. There is this incomprehensible excitement, this dark joy of danger, the horses sink, we have to get across. Only yesterday we had ended up roaring with laughter at the thunder announcing the third hailstorm, joking cheerfully under the waterspouts which immobilized us behind the boulders, lost at this high altitude on soaked rock faces, so close to the bolts of lightning that we could touch them. (In my notebook, that the evening, I write: 'There is a secret law, inherited from Sparta and the Lakota Sioux: in the face of risk, of difficulty, we smile, and then we whistle.')

We finally cross where there is no real danger, and the horses snort. They drink the melted snow in the basins. It's 5 p.m., and Djoki rummages, without turning around, in the package at the back of his saddle, before grabbing a metal cup, white and ornate. We slip along our horses' flanks, and kneel at the foot of the water source which surges up under the névé. The cup fills with this icy water, freshly poured from the sky's hands, and we drink in turn, in silence, gazing into each other's narrowed and shining eyes, gently nodding our heads.

(Darwin wrote in his notebook M a sentence that comes to mind: 'cold water brings on suddenly in head [sic], a frame of mind, analogous to these feelings, which may be considered as truly spiritual.')[3]

Then we set off again. In my notebook I write, tossed about by the horse's steps: '6 p.m. Pure joy of descending the emerald valley at dusk, the sun has almost set and its light smooths each ridge, sketches each flower, the rising wind washes away all thought, and the joy of the quickening horse, transmitted by the saddle, from the pelvis to the soul.' It is fatigue that finally melts the two bodies and the two

The patience of the panther

perspectives of horse and rider, thus accomplishing, without will, and almost without our agreement, the quest for 'centaurisation', as an author I have forgotten puts it – this is perhaps the mythical foundation of horse riding.

We arrive at the last pass before making our descent to the hut. It's 7.34 p.m. We go to bed in the half-light. At this hour the large red deer will come up towards us. Djoki knows their habits; we hide to watch them pass. The pass is covered with heather, the sun sets on the horses, whose tails tremble with mingled pleasure and fatigue.

Here, we no longer do things by listening to ourselves, to our little inner rhythms. It's raining, we're not going back, we continue to explore this trail as if nothing had happened. We eat when we have time, between two surveys: it's all about letting the rhythms of other living beings guide the decisions of the day.

Then, suddenly a red deer crosses the ridge up there, a large male with antlers branching out like a forest! He's sensed us, he disappears. To horse! We track it at full gallop through the valleys, the horses, invigorated as they catch our excitement, prance in the heather, our eyes are riveted on the tracks, we absolutely mustn't lose his track. He's vanished. The horses paw and snort, looking for a reason to keep galloping.

During its evolution, the wild horse has established flight as a defence strategy against predators, which used to include us. Taking flight, according to some people, is this horse's reaction to everything, his existential tactic, turned into a source of power, a necessity transformed into art, hence his joy in running. From this point of view, according to the master riders, the art of horseback riding is to stop him taking flight ahead of you, teaching him instead to take flight with you on his back. And in a direction that you have chosen. Now we are tugging the reins towards the Naryn River and the camp.

The patience of the panther

It's 8.07 p.m. Exhaustion. The horses have spotted the cabin at the bottom of the valley, they descend the steep paths in a dazed, disorderly trot. In my notebook, I write: 'To be nothing more than a bag of meat perched on a mechanical bull.'

In the evening, we sit cross-legged on the soft horse blankets, placed in a circle on the grass, which has barely recovered from the hailstorm. In an exquisite metal teapot, as courteous as high-society ladies, Djoki and Mairenbek make and stir the tea boiling under the black sky, in the deep copses, that sweet tea which makes a home in the hollow of every out-of-doors.

It's like a typical day in the life of a Naryn ranger. A Kyrgyz proverb says: 'If you exist, be like the Kokh Zal (the chief of the wolves, the bravest).'

Okay. But tomorrow morning.

Intermingled days

The days pass, along the river. Our caravan climbs up towards the high steppe plateaus and the tributary of the Naryn called Djeungueureumeu ('rumble'), towards rock walls where the rangers have already observed the panther several times. Ten days of expedition already, surveying this closed reserve, between 2,500 and 4,000 metres above sea level. No other human beings, no screens; horses, smells, riverbends, hillsides, thunderstorms and heat waves; weather which passes across the sky as if in fast forward. In my notebook I write: 'To be each time just where the body wants to be, to squat down on a clump, to lie down in the grass, everywhere at home and always in the home of so many others, a shared, tangled home that can't be owned. To be as dirty as the earth, the forest, the meadows – that is, as perfectly clean as they are.'

The patience of the panther

This morning, we are investigating an intriguing area: an interlacing of rocks that have slid down the gentle slope of a meadow, creating a network of galleries and caves. The harvest of droppings is stunning: more than seven species share this place. Is this a refuge from the storm, a meeting place, a hunting reserve, a haven of peace, a non-aggression zone? For the clues left by prey and predators coexist on the floors of the *same* caves: panthers, bears, wolves, foxes, marmots, martens, pikas, ibex and Pallas's cats . . . (There are certainly risks of error in identification, we often make mistakes: that's one of the charms of tracking.)

We call the place the 'wild agora', and of course set up our mechanical eye, a perfect, mindless memory of everything that takes place: the camera trap. The riddles of this place deserve to be taken seriously.

We set up camp along the river, under a herd of placid ibex, sheltering in the rocks of the cliff above. We take turns watching them through field glasses while the others prepare the *plov*, a stew of vegetables, rice and fatty mutton.

I write in my notebook: 'Today I found my first animal scratch and an almost evanescent panther hair! So light, barely visible, barely hanging from a rock wall'. If you make a little effort to see the world from the perspective of a body other than your own, you eventually find a fragile fluff in the midst of a rocky chaos. You sense the passages that draw the panther's eye and her panther body, with her own style of movement, the places she will want to mark, the perfect rocks. The panther marks the walls by rubbing herself on them, or by aiming at them a jet of urine rich in meaning: her desire is that these coats of arms and flags that express her identity and the boundaries of her territory can last as long as possible in a weather-beaten world. So she will choose rocks that are *sloping* with an overhang that protects the markings from rain and winds. We end up almost being able to guess,

The patience of the panther

from afar, which rock may have attracted her geopolitical desire to mark her territory. Yes, we could almost end up seeing the invisible, seeing how the others live, those pure ghosts, and following their trail.

Here my panther has scratched the earth with her large forelegs, to deposit signals, and I can follow her trail from this point. Just a little higher, invisible fur-covered paws have crushed the grasses in her descent, she has slipped by here, I can see her, her body gracefully negotiating the bend on the ledge, and so her next pawprints will be right here; and higher, in the sinuous axis which summons her to cross the rocky chaos, I find another scratch: she passed this way heading *downwards*, as her weight has made a definite impression through her broad front paws, we can tell this from her prints. We follow her, we go up a line of her past like a river of memories, a river of tiny signatures. Without seeing her, we have entered for an instant the panther's world, her surroundings, perhaps more profoundly than if there had been a face-to-face meeting.

What is demanding and fascinating about tracking is its utter lack of the spectacular. To track the beast here means looking for a panther's hair in a glacial valley. It's all about seeing the invisible, conjuring up intangible habitats in an enriched world, where we have never been alone.

Tracking consists of being attentive to the network of invisible influences that structures the visible living world, making them stand out, by surveying them – and not in collecting name-tags for each species that one encounters.

Wolves, panthers, bears remain invisible – it doesn't matter, we track them among the ridges: better than seeing them with our eyes, we see through *their* eyes, we approach their world from their perspective, as much as our plasticity of living-like-them allows; roaming this shared world, this common, tangled realm.

The patience of the panther

Tracking turns us into a multiplicity of sensory animals: scrutinizing each tuft of grass like a super field mouse, and a moment later, embracing the endless line of the ridge like an eagle. Being able to distinguish between the hair of the groundhog and that of the ibex in a dropping, and then counting in the azure sky the underwing spots of the Himalayan vultures. It's a gymnastics of the crystalline lens, and of the mind: to feel and live on every scale, from the blade of grass to the sky god, and back.

Every day here makes us become an animal of a different kind, an animal that does not forget it's an animal. An animal less mediated than usual by concentric layers of technical envelopes, an animal with an intensified sensory life and a more sensual spirituality: cosmic gratitude for being alive, every morning a sunlit handshake, purification by the frozen spring water. Our gratitude goes to the rich meat that gives itself to our mouths, and not to some abstract being that supposedly created that meat. But neither is it the lack of gratitude found among the moderns, as if this meat were their due, or a pure product. For this lamb came from a sheep, from this sun and this grass, here right under our eyes. A more forgetful animal too.

In my notebook, I write: 'The wild garlic that stings the mouth, snatched by the hand when we go to look for the horses. The tangy rhubarb that we pick without getting off our horses and that quenches our thirst. The pollen that the horse's flanks scatter, sending seeds ever further. It's the great circulation of the sun-flesh, which is the basis of our sustenance. Relearn to give things back.'

To track, in the broadest sense, is to read all the signs (and there are quite enough signs of living beings, we don't need to look for signs of destiny or the future).

To seek, desire, feel, see, understand everything: this attitude often projects everyone outside of themselves, into an enlarged self which no longer has so many cumbersome subjective problems, but makes room for other living beings. In this great spring cleaning of the self, the world living outside finally moves in, into the hollows of the interior ecosystem of links and affiliations. And the other living ones settle under your skin (don't rush, everyone will have their place).

By observing for a long time, out of the corner of my eye, the rangers and our guide, I realize that this practice of nature is above all a decentring of a rare type: to be as if thrown out of oneself by breaking away, bouncing from the springboard of the binoculars out of one's field, far from the self and one's fellows. The binoculars are an instrument of spiritual exercise: sharpening the hawk-gaze with such intensity, projecting far from oneself a focused attention to the other and a headless openness, where the ego almost disappears in spite of itself (I challenge anyone to think about himself when pointing binoculars).

It is an instrument like any other for 'working towards one's own disappearance'. But this loss of the ego does not come with a sacrifice of the ego, with any guilt over being oneself. It is forgetting yourself in the sense of forgetting your umbrella, you're so captivated by something else. We leave our ego on the coat hanger, because the world and other selves are, for once, more interesting than the ego.

This philosophical tracking that I think I can sense in the rangers is the attitude of a living person with an incandescent and disinterested interest in the living creatures before him. A living creature fascinated by living creatures, but *among them* – one who feels that we are living creatures before we are humans. One who seeks out the common in difference, the common segments that form the basis of our particular animality: our human way of being alive.

This forgetting of oneself in the quest for the other, as a spiritual

The patience of the panther

technique of decentring and expanding the self, produces an effect that is at first glance paradoxical: it is *human* relationships that an interest in other living things enhances. How does the wild make us better humans? It's a mystery. Yet the rangers, Djoki, just like Bastien, our extraordinary guide and leading naturalist, share this paradox of being so human because they so much love non-human others. Attention to others, decentred openness, a benevolence not tied down to its own inner rhythms and demands, give a strange quality to our human relationships in the middle of the mountains. A group of naturalists forms, by a strange loop, a humanist circle. There is also the attitude shared perhaps by all *coureurs des bois*: a serene acceptance that it is 'nature that decides'. It implies a relationship with the living world in which no one considers that everything is *owed* him, and this relationship surreptitiously affects relationships between humans.

In the evening, in my notebook, I write: 'To be cold, to eat a hot meal, to find the comfort of being homogeneous to everything, the paths are welcoming to the feet, the smallest clod is a throne, every stream a fountain, each cloud-dodging sun is a caress, and each wind a caress, and if it's cold, that's good too, and if the lips crack like the pale ochre cliff in front of my eyes, that's good too. Your hands are brown and weathered on their backs, where they soak up the sun and the wind when you hold the reins, and white and tender inside, like almonds – as we are now.'

At nightfall, I hear Makou singing Kyrgyz ballads to buck himself up so he can walk up to the horses. The human animal invented song in the encounter of the natural rhythm of walking, the repetitive strike of the striker on the flint, and boredom, like a sail that carries you further while swelling your heart.

The patience of the panther

The virtue of patience

Only one more day before leaving the reserve and the River Naryn.

This morning again we stalked with our eyes, on the ridges, the traces of the panther, and its silhouette which might be drawn on the sky, erect, our binoculars always ready to rise, in an arc of a circle, from the chest to the face, the same way you put on a mask.

A falcon mask, which gives us the superpowerful sight of the raptor, and a supernatural intimacy with these unknown creatures. I am thinking of an expression used by the anthropologist Eduardo Viveiros de Castro who sets out to define the meaning of metamorphosis in animist traditions: it's not a matter of slipping an animal appearance onto a human essence, but of 'activating in oneself the powers of a different body.' For example, the evolutionary power to distinguish, a thousand paces away, the silky detail of a plumage, the calm expression of an animal gaze. Animist metamorphosis is like adopting another body's perspective on the world around it. Binoculars, then, are in a sense a transformative mask: a Horus mask that transforms us, from within, into a were-hawk, for the time it takes to pierce the curse of distance with the focus of our eyes.

I see an animal far away, a living comma at the end of the binoculars, undeniably alive, although almost invisible in the rocky chaos. He disappears.

In my notebook, I write: 'Seeking to see without seeing. The joy of peering. The joy of searching without finding, the joy of freeing oneself. The wait, the imminence, the ardent patience. Seeking, seeking to see, in vain, the mottled exterior of the snow leopard, is beneficial to man's inner life.'

Seeking to see the panther for hours through binoculars, scraping your eyes on every ridge and rock, letting the tears of bedazzlement

The patience of the panther

trickle at length down the cheeks behind the lens, because we've forgotten to blink. Just seeking to see her, with such ardent interest and in such a selfless manner, just to place our gaze upon her like a wing on the wind, without leaving a trace – but what sustains that endless desire within us, that patient desire to see the panther?

We lie down in the grass again, without a word, in a stable position, to improve the pressure of the binoculars as we observe, and the quest resumes. The density of attention is almost solid around us, the sky passes heedlessly over us, who are all gaze.

Suddenly, I understand in all its clarity that it is with the patience of the panther, with her patience, that I am seeking the panther.

I risk this conjecture: this strange aptitude of bringing a loving patience to your quest, that ardent patience, that intense mastery of attention, is an animal ancestrality deposited within us. A legacy from that phase of our primate past some two million years ago, when we changed from fruit pickers into partially carnivorous trackers. It is with the patience of the panther that you track the panther. This is not a metaphor; it is a shared animal ancestrality. This phenomenon resembles what biologists call evolutionary convergence: a skill segment shared between several species due to a momentarily similar evolutionary history. Convergence is a concept in evolutionary theory that characterizes traits that are profoundly alike in two species, even though their common ancestor did not possess them. For example, the hydrodynamic fins of dolphins (mammals) and sharks (cartilaginous fish) are profoundly analogous because they have experienced very similar pressures of selection over long periods of evolution. The most cutting-edge research is now investigating the possibility of applying this concept not only to the organs of the body but also to behaviour.

The patience of the panther

Based on this notion, we can build the idea of the 'animal ancestralities of humans', seeking them out within ourselves, in our sedimented heritage, to better understand who we are.[4] The panther's patience, as a behavioural skill, seems to be part of those cognitive and emotional matrices that we share with certain living beings, those with whom we have shared similar ecological living conditions during a long phase of evolution. For it is with this same ardent patience that all living beings who have evolved in a world where they must spot and capture the life that will nourish their lives scrutinize their surroundings, hunters on the lookout ready to move in. With analogous ecological needs come analogous pressures of selection and therefore analogous behavioural solutions. When there is an advantage in being a patient, focused, observer of your prey, rich in controlled desires, this behavioural trait fits into your dynasty of living beings through natural selection. This patience is thus a gift of eco-evolution.

On this view, panthers and human beings inherited this special form of patience at the same time: of course, we do not inherit it from the panther itself, as she is not our ancestor. This is the meaning of the concept of behavioural convergence. It is just for convenience that I attribute it to the panther, because it is in the externalized medium embodied by the panther that this type of patience is best seen, in a purer way than in our own human behavioural matrices, all tangled and intertwined. Like a certain roll of the eco-evolutionary dice, the panther dramatically shows some things that we share with her more indistinctly.

It is said that patience is a characteristic of man. Saint Augustine, questioning himself on the source of patience, relates that he mainly takes pride in believing that it comes 'from the forces that the human will draws from the depths of its freedom',[5] finally recognizing that it is necessary to attribute it to the grace of God: 'the virtue which bears the name of patience is a very great gift of divine munificence.' When he sets out to define it ('If we hope for what we cannot yet see, we wait for it patiently'), his formula applies really well to every predator.

If we agree to investigate the origin of patience, not on the scale of a few millennia of civilization, but on the scale of millions of years of our evolution; if we agree to draw up its Nietzschean genealogy on the scale of geological times, the origin of patience appears in another clarity: it could in fact come from the panther that we were – that is to say from our primate ancestor who, some 2.4 million years ago, had to learn to hunt, unlike all of our primate cousins who remained frugivorous. If you observe them well, these relatives never scrutinize *other* species with such duration, such curiosity, such intensity; they are obsessed by their own interminable family sagas: it is because they have not known the pressures of powerful selection that bear on the characteristics of the stalking animal.

It would thus be the evolutionary period of tracking, hunting, stalking and approaching that brought about the very particular type of panther patience in the human line. A way of life that lasted more than two million years must have left its mark on the way we act as living creatures. In this sense, it is because we are panthers that we yearn for the panther. This patience has nothing to do with killing: it is fascinated by other forms of life, and easily turns away from its origin in hunting, to a selfless interest, like that of the naturalist. Like the interest shown by anyone who, returning home, is hypnotized for a moment by the beauty of an alert deer caught in our headlights.

We have to be precise to break away from the old myths that tell us who we are, we have to compare.

This ardent patience does indeed differ from that of the scavenger, which is a mere phlegmatic expectation, but which also probably exists in part in humans (a few hundred thousand years of scavenging must have deposited in us a few behavioural matrices adjusted to this practice). Mouflons, which have been observed for a long time, do not have the same attentional determination towards other species either. The patience of the ibex looking for a predator does not have the same texture, nor the same emotional tone: it is vigilance in the face of possible danger, it is not veined with desire.

I think of the wolves I saw observing a herd of elks in Montana. The sated wolves who, without hunger, scrutinize them, fascinated. With the same ardour as ours, the same selfless joy at being sated. Evolution gives you a joyful and immoderate interest in what is good for you. We can see the Katmai River in Alaska, thanks to trap cameras connected to the Internet, and watch a bear waiting on a promontory in the middle of the torrent, so patient, so diligent, motionless, all senses on the lookout, a delicate colossus; we can clearly see that this desiring patience is an animal virtue.[6] An animal virtue that we have also inherited, and then diverted into such strange activities as listening to the teacher on the school bench, like brainwashed bear cubs.

The animal ancestralities of humans

So it would be precisely with the patience of the panther when she is *sated* that we track her. Stripped now of all desire to catch and eat, through changes in living conditions which are different conditions of expression of the same ancestrality, which here combine it with others.

Our current life as trackers 'in a world without prey' facilitates a reinvention and reassignment of the cognitive and emotional matrices drawn from the lives of our ancestors, adapting them towards

The patience of the panther

a thousand improbable things. Diverted from other animals to any other medium. It is the ardent patience which drives the naturalist and the wildlife photographer, which is active in any investigator, researcher, bargain hunter, explorer of the streets, of books, of the Web. We have exapted this ardent patience, in other words we have *diverted* it towards so many other objects of desire. This patience stems from our hunter-gatherer past, forged by nearly two million years of evolution, now available for a thousand purposes still to be invented. It's not only when we are tracking that we activate it in ourselves – but it's because we have been tracking in the distant past.

Of course, we also have other forms of patience, we were fruit-eaters and gatherers long before we became interested in other species: we have, bubbling away within us, a thousand sedimented animal ancestralities, from a long past; but they are not expressed under the same conditions. The patience of the panther does not exhaust all forms of human patience. Remaining calm in the face of the person who infuriates you is the patience of the baboon, if we are to believe the hypothesis of primatologist Shirley Strum who sees in the play of the young baboons that she observes an exercise in self-control. The patience of parents, that of remaining calm in front of a child who exhausts them with demands, is not the same patience: this is the patience of the wolf, as of any social species where parents contribute together to bringing up their little ones.

The mode of subsistence constituted of picking and gathering has certainly also deposited in us, sedimented, other animal ancestralities, behavioural matrices that we will understand better by looking at other living beings. They are probably in some ways more intense than tracking for we were leaf- and fruit-eating gatherers earlier and for a longer time than we were trackers, hunters or scavengers.

The patience of the panther

Once again, experimental protocol means that you turn the animal that you are into your own laboratory, you have to go out into the field and pick wild plants, berries and nuts for a long time in order to see rising back within yourself, far from our supposedly properly human oddities, the behavioural matrices shared by other gathering animals (and only excellent specialists in wild gathering, like François Couplan for example, would be able to formulate precisely, with all the delicacy of someone like Henry James, the fine architecture of this matrix).

The vibrant attention that becomes active when we seek the type of meadow that can accommodate the goosefoot or the sour sheep sorrel, the recurrence of the link between the dead trunks of Pyrenean pines and the wild strawberries which seem to enjoy thriving near them, exposure to the sun and the altitudes where blueberries swarm; the joy of searching, classifying, distinguishing, ordering plants in the mind are probably in part inherited from behavioural matrices from the gathering animal that we once were.

Because we have been gatherers for much longer than hunters, we have inherited another patience from that time, the patience of the animal gatherer who tirelessly strolls, chooses and delicately collects a thousand differentiated plants, each for his own use. It is the nonchalant patience of the deer, which chooses its favourite grasses and even knows the particular barks that cure its stomach aches in spring.

It is a demanding herbivore ancestrality that has become sedimented within us. The traditional naturalist, again, sees in this his specificity, the rational, abstract and disinterested human spirit – as an ungrateful heir of skills and forms of attention which are an animal heritage. He has exapted them, that is to say, diverted them from their original function, to practise the botanical art of taxonomy, but it is to his ancestors that he owes his gratitude for being able to do this, and

The patience of the panther

to enjoy doing it. Because the behavioural matrices encapsulate in us a mixture of powers and desires.

A minute of meditation here, in a strange, new, necessary type of ancestor worship – the worship of our prehuman ancestors.

I hypothesize that all visual, vibrant, loving patience, which tracks down what it seeks, which knows how to wait, and in which while waiting exacerbates desire rather than snuffing it out, is analogous to the panther's patience: only animals possess it which in their evolutionary history have known this behavioural convergence (just look at the body of the cat which trembles with all its muscles under the opposite effect of the desire to leap and the discipline of waiting, a pure self-control which alone will ensure the vital success of the leap).

If we accept that the behavioural patterns of the living beings all around us have been subject to evolution, there must be in us, not just in our bodies, but even in our minds, traces of who we were. These traces are multiple, they combine together, open up a thousand possibilities for us, they form the basis of the intensity and plurality of our desires, the obviousness of our affects and the structured inventiveness of our ends. They are the palette; each individual is the finished painting.

If we are so free, then, it's not that we don't have 'instincts': it's that we have too many of them, they bubble away and recompose themselves indefinitely. These animal ancestralities are capable of forming new affects, desires and temperaments, since from the oldest to the most recent they *simultaneously* lie 'open on the surface' of our experience, and are therefore able to be combined. We are rich in a thousand internal animalities, like other animals elsewhere, but our cultural and technical way of life allows us to combine them in a plural way – to reveal their multiplicity, to express them in endlessly new constellations. Institutions, customs, technical systems balance these ancestralities within us in different ways. The technical niches that are our cultures modulate in countless expressions our inherited animalities.

We can recognize the founding vectors of our own identity in the playing ibex, in the male eagle when he makes an offering to his companion for her to choose, in the panther hypnotized by the object of desire, in the curious wolf out on patrol, exploring new worlds as he disperses them. In the bear, that tireless taster. The animal ancestralities of the human here are something like the embodied figures of the ministering spirits of animist traditions – scientifically documented.

Practising the panther's patience thus invents another way of being a naturalist: one who practises his art of observing animals without forgetting to be an animal. One who knows that his art is an art woven from animal powers which he finds and reactivates within himself, which he finds outside of himself. It is a gift of eco-evolution that this panther's patience, this joy, an ebullient energy and an intensity of research that leaves its mark everywhere else in life, and is truly life's salt. The reversal is strange: not the curse of being an animal, but gratitude for it, carried by vital vectors inherited from evolution, vectors that are so powerful, so capable of fulfilling us, so easy to emancipate and divert towards other ends (because we have not finished inventing the research, the work, the projects to which humans will devote themselves, with their endless panther patience). The possibility of emancipating our animal powers for unheard-of ends is also a gift of evolution, of its original plasticity which promotes exaptations in all living beings, those unforeseen changes of the use or function of our biological inheritance.[7]

The patience of the panther

The animal art of living

The panther's patience is not a moral virtue of elevation above the archaic and primal animal that we imagine to be lurking in our inner hearts; it is not a muzzle placed by Reason on our supposedly bestial passions. To practise this patience comes down to 'activating in oneself the powers of a different body', as anthropologist Eduardo Viveiros de Castro wrote to define metamorphosis.[8] This constitutes a metamorphosis in the animist sense, but one enriched with an eco-evolutionary understanding of those animal powers that we bear *within ourselves*.

For patience understood as the mastery of one's impulses and of one's inner discourse, a patience conceived in all philosophy, from Zen to Stoicism, as the ground of a wisdom that seeks, is probably in a sense found in higher degrees in the lynx on the lookout, the fishing bear, the panther on its approach to its prey, than in us. For they have, more spontaneously than most of us, the freedom from parasitic thoughts (which is also their powerlessness). We have to acquire it. We humans are assailed, whenever we have to be fully present, by that endless inner discourse that moves away from mindfulness. Contemporary attention-catching devices make it impossible for us to fix our concentration on a present or a sustained desire, to keep pace with things.

Only the Western thinker is weird enough to believe that wisdom involves moving away from the animal within, rising above its rubble: the Zen sage, it seems to me, seeks to get closer to his cat – to its own powers of wisdom. (I also think of the chaffinch singing in the morning in the nesting box on the terrace, making fun of me for already thinking about tomorrow, about next year.)

There may be an ethology of animal wisdom to be written. The

The patience of the panther

panther's wisdom, for example, extends beyond its ardent patience. The quiet sovereignty of the solitary feline, his ability to inhibit alienating associative thought and to enjoy the tiniest gifts of the living world around him, make him a master of domestic wisdom who can probably inspire the wisest of us, by inventing a manner of life that human beings, social primates dedicated to games of power and influence, have not been able to invent alone. For evolutionary reasons, the solitary feline has invented an ethological life form which is a form of sovereignty without subjects. Or: how to be a king without having power over anyone, that is the paradox – without owning anyone, and so without being possessed by anyone. Watching the panther move over the crests, or your cat in the living room, it is clear that they have found an ethos of majesty that the great kings of the earth have seldom attained. Their independence is the ethological privilege of kings without kingdoms, unconquerable because they have nothing to lose, impossible to subjugate to our power because they have no power over anyone. The solitary feline is imperial with no empire other than over itself. There is very little weakness in his ethogram, because he has little or no vital need that he cannot fill on his own; and for this reason, his affection is something like a voluntary gift, not an expression of dependence. There is this formula of Nietzsche which is enigmatic as a political ideal: 'There is mastery only between masters.' It becomes obvious when you live with felines.

A thousand other forms of life have strange, often involuntary and non-linguistic forms of wisdom that can be learned.

And yet we are the heirs of a culture which, in broad outline, has seen wisdom as an elevation above the animal, within and outside of ourselves. For this, it was necessary to disfigure the real animal, to erect it as a foil onto which all human vices are projected (the animal

The patience of the panther

is ferocious, bestial, unable to control its violent or sexual impulses, improvident, etc.).

Other heritages are more lucid: certain ancient wisdoms (the wisdom of the Cynics and Sceptics) endeavour to rediscover an animal tranquillity from the time before language – the very name of Cynics becomes less enigmatic in light of this idea. The Native American shaman Davi Kopenawa also has this strange wisdom: he treasures his macaw feathers, because they give him the animal power of eloquence, so that he can speak with the white chiefs in three-piece suits who are destroying the forest.[9]

What we can learn from the sovereignty without subjects of the solitary feline, from the curious dispersal of the crow, from that indefatigable taster the fox, from the concentration without parasitic thoughts of the bear, are animal arts, animal arts of living which, hidden inside us, have always nourished us with their power, arts that we have merely added a twist to, or else forgotten.

It is a logic of difference without the pride of being unique. Between living beings, there is no difference in nature or degree, but a difference in problems to be solved and in evolutionary history. It is a logic of kinship where powers are shared; life inside and out is fascinating, and admittedly other animals cannot solve differential equations, but what does this really matter?

Day 12

It's our departure day. The panther of flesh has remained invisible on the ridges. But we now have it under our skin, it can be activated and help us to live stronger, more vivid and wiser lives. And then, by dint of looking for her from inside her own point of view on the world, we have got to know her, thanks to the delicate knowledge of

The patience of the panther

our guide and the rangers. The traps that we set on what we hoped to be the familiar paths of the snow leopards are running, with their mechanical, but infallible patience, machines that carry our desire to see and understand into our very absence, into the silence and solitude of the glacial valleys.

A few months after our return, we receive an email from the rangers who returned to the panther ridges to check the camera traps at the end of the summer, and sent us the pictures: there they are.

4

The discreet art of tracking

It's August in the Haut-Var, and here we are, a few friends who have gathered to investigate the presence of a wolf in a remote valley.

A whole bunch of information, gleaned from the ever-useful bulletins of the National Hunting and Wildlife Office, Internet forums and village betting shops, leads us to believe that something might be found here. None of us, however, could have expected what we will find. A site so rich in clues to harvest, puzzles to decipher, reciprocal tracking – a site of a kind I have never seen since.

Just upstream of a deep canyon, at the crossroads of two human trails, our noses warn us: there's a sheep's carcass lying on the ground. The bones are scattered here and there. We go around the canyon by the forested western flank, very dense, very steep, navigating with animal eyes, taking alternately wild boar paths and deer tracks, finally emerging into a valley with a strange atmosphere.

On the western flank, dotted with pines and downy oaks, wild boar *boutis* (those areas of land that they turn over with their snouts to dig up the tubers) are omnipresent: we're in their territory. The forest seems untouched, but on closer inspection, everywhere, beneath it, buried between the roots, laid bare by the earthworks of wild boars, dry stone walls tell of an intense human occupation, but one dating back to ancient times and now vanished. On the other side of the river, on the eastern slope, a few meadows are still used to graze cattle, far from it all. The meandering

The discreet art of tracking

river is very clayey, turquoise in colour, almost milky on the shallows.

On the lookout, we uncover a first canine print in the dried clay. We are not experts, only enlightened amateurs, but we are starting to recognize the clues that identify the wolf's trail. The trace is massive, almost eleven centimetres. Diamond-shaped, clearly outlined, strongly clawed, with a large space between the central ball of the foot and the front balls – very lupine. But it is impossible to be sure from a single trace. Our attention sharpens: a solemn, silent, immemorial excitement takes hold of us, a fiery joy that focuses our gaze. Further on, there's a long track that unfolds before us. We tie two belts together to make them run along the pawprints: they are perfectly aligned, in a single thin line – a powerful sign that this is the wolf. The dog, a descendant of the wolf selected for his docility and his ability to attach himself to humans, resembles a wolf kept all life long in a juvenile state: the dog's trail is more muddled, his pawprints draw two parallel lines. He's lost the perfect wolf's art of surveying, which optimizes every movement intended for wandering.

Before our eyes, the clues accumulate: the prints of the hind feet are placed *in* those of the forefeet, more massive, in an almost perfect overlap. This is another sign that distinguishes the two animals, probably an old adaptation that the dog has lost and which allows the wolf, a sovereign hunter in winter, to run through the snow while minimizing his efforts. The animal we are following here is probably moving at a fast trot, as the hind pawprint protrudes slightly ahead of the front one. We are following in his footsteps.

When we arrive at the river, where a path crosses it and makes a clearing, we are stopped by the spectacle. Dozens of traces dot the clay. There is no human trace, either here or on the path. All the traces are of *Canis lupus*, but they are of different sizes. A set of mas-

The discreet art of tracking

sive pawprints, probably a large male, two other different medium-sized tracks (female and a young adult?), and one last, very small one – a yearling wolf cub or she-wolf.

We go upriver in silence, in the hope that, on the basis of a misunderstanding, we will surprise those inhabitants who seem to spend a little of their summer in the hollow of the valley. On either side of the river, a strip of clay one to two metres wide encircles it; the drought has reduced the flow. And on these two blank pages a few hundred metres long, a whole past is written, a series of daily habits, a whole mysterious novel about the life of a family.

Tracking is sometimes like coming home at night, to a very large house with the sky as a ceiling, and following through the rooms the omnipresent traces left by the beings who live with us here; the poignant little clues of their daily activity, everything they did while we were away: the bowl left out on the kitchen table with its easy to forgive leftovers of cereal; here, slippers abandoned outside the shower; all those little clues where you can track the activity of your loved ones, their concerns, right down to their state of mind. Their art of inhabiting, and of cohabiting with us, entangled with us, in this common world.

Before our eyes, hundreds of pawprints intersect in many copies in the clay. They live here. We set up a discreet camp a little further on. One enigma begins to emerge. Why do these sets of tracks run along the river for tens of metres? At this time of year, the cubs have come out of the den. The entire pack focuses its activity on a specific site in its immense territory, in such a way that this is the only time, after giving birth, when it settles down around the litter. This place is often a somewhat protected clearing; wolf specialists call it a 'meeting place'. Cubs are stationed there, often guarded by an older family member, a big brother, an aunt, father or mother, while the rest of the pack goes hunting. Sated hunters return daily to regurgitate up to seven kilograms of meat for the greedy cubs. It is common for the meeting place to be a few hundred metres from a stream or river to quench the thirst of the little ones and the pack. At this time of year, one method of finding packs is to go up rivers to look for tracks that cross them perpendicularly: if these tracks are dense, then we can hope that this is where the path from the river to the meeting place leads. But our pack, the family that has made such a good choice in this valley, does not cross the river perpendicularly. It follows it, walks along it length-wise, meticulously. Yet the ground is clay, slippery, stones dot the ground, it must be difficult to walk on it. Why stay here? Why persist in walking for so long through the clay? Even more mysteriously, at several points, the tracks along the river end up advancing towards the water, *perpendicular to the current*, for a metre or two, and stop. Then the tracks tell us that the animal returns to the bank.

But what's going on here?

An art of thinking

Or a game of traces. A track in the clay soil of a river. Of course, there is not much visible: only *Canis lupus* prints in the mud. But with other eyes, we can recompose a trajectory, extrapolate a course, a gait, a bundle of intentions that speak of a certain way of inhabiting a place. The emotion comes down to seeing *with his eyes*; to follow his trail, we are obliged to *move around in his skull* to understand his intentions, to walk with his legs to understand his movement. Here, we see the forelegs sunk, parallel in the mud: he stopped, right where we are, to observe the landscape, and inhale the aroma of the sheep grazing

The discreet art of tracking

below. There, he sovereignly inspected his kingdom; here a scratch (*grattis*), to signify to another pack the border that they will not be able to cross without having to fight, or to size up the tenants. This philosophical tracking was described by the lieutenant of the royal hunts Georges Le Roy, a friend of the encyclopaedists Diderot and d'Alembert, in his *Lettres sur les animaux* (*Letters on Animals*):

> The hunter, by following the animal's footsteps, seeks only to discover the site of his lair; but the philosopher reads them as the history of his thoughts; he disentangles his anxieties, his fears, his hopes; he sees the reasons that have made him proceed with caution, those which have led him to pause or speed up; and these reasons are certain, for otherwise, as I have said, one would have to assume effects without a cause.[1]

In his field investigations concerning the tracking practices of the Bushmen hunter-gatherers of the Kalahari, anthropologist Louis Liebenberg formulates a hypothesis on the role of tracking in the emergence of human reasoning skills (this idea will be developed in the last chapter).

The hypothesis is the following: the human being has developed intellectually from the point of view of aptitudes to decipher, interpret, surmise, because nearly three million years ago he moved in an ecological niche where to find your food needed *investigation*. Native hunting animals often have a strong sense of smell. The whole problem comes down to the fact that we were originally endowed with the bodies of frugivorous primates, that is to say *visual animals without a sense of smell*, who subsequently became hunters and trackers, that is to say, doomed to find things that are *absent*.

For this, deprived of noses, we had to awaken the eye that sees the invisible, the eye of the mind. In tracking, we see the potential elevation within living things of decisive intellectual skills, which revolve around the power to see the invisible, for example the destination of the animal or a sequence of events from its past. Tracking is an intellectual problem that probably helped to create the human being. Sherlock Holmes is just an extreme form of the primate tracker, our ancestor. When Paul Klee says that art 'makes visible', he is another such form, less analytical but just as hypersensitive to the cosmos of clues.

In our valley, the puzzles deepen. One particularly impressive set of tracks shows the pawprints of a massive animal, possibly the breeding male (the one scientists used to call the 'alpha male'), followed on his heels by the pawprints of the cub. The large male's pawprints then sink into the river, and the cub follows him, but only up to the water's edge.

Even more intriguing, the canyon just above is occupied during the day by canyoning enthusiasts. These strange bipeds are decked out in fluorescent suits and helmets, and they shout 'banzai!' as they jump into the pools. On the other hand, the lower part of the valley, closed off where the track has collapsed after a flood, is much quieter. Yet the wolf tracks are focused here, right here, very close to these humans, glued to their activities. Why does the pack not go further, to where it's calmer? The mystery remains unsolved.

In the evening, we bathe not far from the tracks, before going discreetly to grill our meal on a fire, well secured by buttresses of damp clay.

The next day, however, a strange, hitherto insignificant gleam of strangeness catches our eye. River crayfish can be seen here and there in the shallow water just downstream of the canyon. The

The discreet art of tracking

hypothesis emerges alone from the obscure workings of the animal brain, without being summoned: what if they were fishing? The mind resists, because it seems unlikely for our 'top predators', but then a memory from my reading comes back to me, reminding me of wolves in northern Canada who fish for salmon and maybe crayfish. The physiognomy of their trajectories here, however, finally makes sense: the cul-de-sacs where they advance into the water could in fact be fishing positions. The tracks that meticulously follow the river would then be patient sessions in which they seek their prey. Our eyes now know what to look for: what might we find if this strange story we are telling ourselves were true? This is the essence of tracking: the past is invisible, but no one can exist without leaving traces. It's a matter of deducing the *visible* implications of the invisible hypotheses, and looking for them in the landscape – tracking the traces of the past percolating into the present. This is how we seek and find, here and there, some delicate pinkish debris on a bank, which turns out to be the remains of some river crayfish, awkwardly peeled, or rather shredded, half devoured. We will have to keep looking: the hypothesis has become likely and interesting, but is still not absolutely confirmed. Tracking is not an exact science: it's an action-science where each hypothesis directs our steps and gaze elsewhere – where it spurs our desire, not to conclude, but to seek further.

If this hypothesis is true, however, it teaches us a lesson: we thought wolves would spontaneously flee human activity in the canyon. Our species knows what narcissists we are – we thought we were the main cause for their behaviour. But if our hypothesis holds, things appear in a different light: in the final analysis, they didn't really care whether they were near or far from humans; they were there to fish for crayfish, and this is where the crayfish are, at the

The discreet art of tracking

bottom of the canyon. From the perspective of the wolf himself, who is interested in things that we can't see, his way of inhabiting a place can make sense, its paradoxes can become almost transparent. Wild animals are not, for the most part, confined to a wilderness: they live among us, as interstitial animals. But it is by themselves that they are among us, following their own logics, their way of inhabiting and creating a territory.

The track of the big male and the cub would then tell another story: what if we were being given a lesson about the nature of things? By following these two entwined tracks, we go back in time down their intangible paths, following their own habits in their footsteps. The traces of the wolf cub pausing on the riverbank, his little paws immobilized in the clay perpendicular to the river – this enigma would then be unravelled: he is observing and learning; in front of him, out in the water, his father is showing him how to fish for crayfish; we can still make out the prints of his big paws which advanced into the river and then came back.

This is what we would have found, by going back down a line of their past like a river of memories: the trace of an apprenticeship, of a school day, in this powerfully social species where it is now well established that hunting skills are taught from generation to generation.

The story does not end there. The following winter, I return to the site: no trace. The following summer, I spend a night on the lookout, camouflaged under a canvas with slits like a soldier's tent, in my sleeping bag, along the river, alone, on the lookout: no presence, no pawprints. The following summer, still nothing. A year later, feeling pugnacious, we come back, a whole group of us. We can't find anything on the site itself. I also know that wolves have been shot at in

The discreet art of tracking

this area: has the pack been killed? The absence of pawprints on these same shores, in the same places, leaves a strange, elegiac feeling: as if you were coming back to a holiday home in the countryside where you once bonded with your relatives, but the house is now empty, a home for only memories, and silence.

And suddenly, a little further on the forest track, the trace of a large canine catches our eye. It could be a dog. Patiently, by drawing on everyone's vigilance, by combining our interpretations of the trail, we end up retracing the complex route followed by the animal, from the trail to the bank, for about fifty metres. At one point, the track disappears. By weaving together the clusters of clues, we mentally construct where we might find it: the animal arrived from the forest track, he trotted along quite unfazed, he descended into the riverbed, probably to sniff a sheepskin lying there, he must have continued along this axis; but on the left, there is an expanse of clear sand free of any trace, so he must have passed to the right. And on the right, we find dropping: they are full of hairs and bone fragments, with all the characteristic wolf features. They have been left on each side of a ford which here crosses the river: as so often with the wolf, it's in a central geopolitical place, evident to the eyes and muzzles of all, so that the members of the pack, any rivals, any foxes or dogs, will clearly understand the territorial dimension of the marking.

So they're here, they've returned, or maybe they never left? But maybe it's another pack, or a travelling wolf? By following the trail of the markings, finally, we discover the key element lying on a path, the clue that will confirm a hypothesis that has remained unproven for three years. A massive wolf dropping, lined with hair and bones, but also, when we sift through it, filled with shards of a strange material, which we can't identify until the lightning flash of an idea flickers through our minds. It's a whitish chitin, with tints of pink: it's the remnants of a crayfish shell.

There is a singular emotion in finding at the same time a confirmation of the old hypothesis, and the confirmation, almost the proof, that it's the same individuals who are still prowling round here, after three years of absence: the same pack, because it's endowed with the same tradition, probably very rare among wolves, since it's not been described anywhere else in France as far as I know.

It's a philosophical emotion, because what allows us to identify that it is indeed the same pack over time, the same family, are not biological criteria (their genetic heritage or their physical appearance): it's a shared culture. It is their hunting culture that allows us to identify them despite the passage of time, which serves as a great transformer: territories may have shifted, individuals may have died, the alpha couple may have changed, other wolves may have joined the pack and taken over. All these things are invisible to us; and yet there are things which resist time better than individuals or sovereigns, and these are traditions. The continuity of the pack, what enables us to call it *this* pack, is not visible here over time because of natural genetic kinship, but because of cultural unity. These wolves have never been seen; we do not know any of the individuals in the group, we do not know who they are or how many there are of them, and despite everything, this single clue confers the precise feeling of a recognition, a filiation, a tradition which confirms their existence and their unity. Extracting them from the anonymity of the biological species, where each individual 'wolf' is supposedly explained by the abstract traits of the species (the instincts or the ethogram of *the* wolf), this clue gives them their very particular style, their own pack history on this specific territory.

The density and magnitude of the invisible are unfathomable: we have seen nothing of their bodies, their actions, we know nothing of their existence, and yet, by tracking the tiny visible traces of their invisible lives, we find out something powerfully intimate about them, specific to them, like a rare custom in an isolated village, a discreet way of greeting each other, a sign of the shared belonging of its inhabitants, even after a diaspora. It's them, there's no doubt about it, or their descendants – beings, in short, who share a community of tradition and original know-how. It's through what seemed the least reliable, the most intangible, the most invisible and the least endowed with the permanence of matter, with the repeatability of nature, that the most robust knowledge comes to light. This type of identification is rare in animals. Here the art of the naturalist becomes a little more detailed: we can not only isolate the biological genus *Canis lupus*, but also identify more delicate groups, beings caught up in family traditions: a shared apprenticeship and the transmission of a culture of fishing.

An art of sharing signs

Back to the first tracking session, in the days of the unproven hypothesis, where we follow the myriad traces they left that summer in the mud. We sleep for the second evening in our discreet camp.

At night, around 3 a.m., we howl together with the howling of wolves, hoping, according to an old technique of dialogue, to awaken the desire to howl among the wolf cubs who are less able to control themselves than adults, and can't resist the call, while adults, who probably sense the deception, often don't bother to answer. But if the cubs start to howl then, more often than not, the whole pack follows them in chorus. In Algonquin Park, Ontario, during a tracking trip, an entire pack of wolves from the east answered our call under the

The discreet art of tracking

moon, at length, without aggression, and then fell silent. Tonight we are howling for the wolves of the Var, singing our best wolf song, but only the wind answers us.

On the morning of the last day, as we return to bathe for the last time, one of us freezes in stupefaction. Where, the day before, he had placed his things on a trunk, there's something in the clay that wasn't there before: two massive prints, a wolf's front paws, parallel, positioned in such a way that the only imaginable attitude we can ascribe to the one who left them is that he's holding his muzzle straight and examining and inspecting the smells left on the trunk by the belongings of us strangers. We then discover, further on, other circumspect traces: new pawprints, on trails that we had inspected the day before. Pawprints dating from last night. In silence, everyone draws the conclusion: we have been tracked by those we were tracking. They're here, they're really here, almost as curious about us as we are about them.

This inversion suggests that tracking does not establish a position of transcendence of the human over living beings, the position of an unread reader, the sole consciousness interpreting a living creature blind to itself. When you go tracking along trails, you become *trackable*.

Often, bent double over a trail, the cry of a raptor makes the tracker look up. He searches the edge of the woods in vain, caught in the circular paradox of the hunt: who is looking at you when you scrutinize a pawprint? Of whose amused gaze are you the carefree object, that is to say the prey? The objectifying relationship with respect to the living is stealthily reversed in the heart of the forest.

To grasp this strange reversal, the first realization comes down to understanding that we are sending out signs. This is what demands

The discreet art of tracking

to disappear. Becoming a sage bush. Rubbing yourself with fragrant leaves, hunkering down, sitting quite high up on the hillside, above the valley, and waiting, until for the rest of the world you're nothing but one sage bush among others. Then, something may happen.

The second realization comes down to grasping that you are being read. One experience from a tracking trip in Yellowstone National Park will indicate the full extent of this life of signs in the animal world. In the Lamar Valley, along the Yellowstone River, I undertake an exploration in an area expressly described by the rangers as occupied by bears. Two buffalo carcasses dot the plain, and the powerful omnivores compete aggressively for them. The walk takes me up a ridge, downwind, so at every turn I might take a grizzly bear by surprise. A pronghorn antelope is walking in front of me. I try to reassure her, to get her to advance a few hundred paces in front so that she can be my scout: her sense of smell is so superior to mine that I'll be able to read from her behaviour, her ears, her tension, her run, the presence of the grizzly that I can't smell alone. But the antelope follows her own sweet path. Further on, I work out where the cries of crows are coming from, triangulate the flight zones of scavengers so as not to stumble across a carcass and the creature defending it. This is when I ask myself: what do other animals read in my behaviour? I decipher from their attitude what they know about the world around them, but don't they do the same? All those long periods in the sage bushes when a pronghorn stared at me, a bison studied me, a black bear stood and examined me, take on another dimension. I thought they were interested in me – were they actually reading something they were interested in about *creatures other than me*? Bernd Heinrich, an ornithologist specializing in corvids, notes that wolves and bears read the presence of their prey from the cawing of crows. An ecosystem in the informational sense of the term can be thought of as a circuit of cross-referencing signs and shared information.

To live is to be generous in signs. It is to give out signs to everyone, against your will, without wanting to – signs that cannot be appropriated. And this is the phenomenological definition of a pure gift. Giving and receiving signs, exchanging them, is the foundation and nature of the great vital politics that weaves living beings together in the ecological community. The practice of tracking appears to be a symmetrical practice from a geopolitical point of view: it is not only a matter of reading signs, but at the same time of being read by others.

An art of self-transformation

If you follow his trail, you can see through the eyes of another: if you take a closer look, it almost comes down to magic, or to those metamorphoses that occur in shamanic rituals where the shaman manages to move his spirit into the body of an animal. As Louis Liebenberg puts it:

> Tracking involves an intense concentration that results from the subjective experience of projecting oneself into the animal. The tracks indicate that the animal is starting to get tired: its stride becomes smaller, it shifts more sand, and the distances between resting places become shorter. When you track an animal, you have to try to think like an animal so as to predict where it will go. By looking at its tracks, you can visualize the animal's movement. What is perhaps most remarkable about tracking an animal and projecting yourself into the animal is that sometimes you feel that you have become the animal – it's as if you could feel the movements of the animal's body in your own body.[2]

The discreet art of tracking

This aptitude plays a decisive role in the transformation of our relationship with the living world. It demands that we pay attention to what non-naturalistic, non-Western modes of understanding the world can teach us about our commerce with other species.

Among animist peoples, the shaman is the specialist in understanding and negotiating with non-humans, especially animals. But to negotiate, one has to pass from one species to another, and this can't be done spontaneously and effortlessly, as the gap between the forms of life implies a change of *perspective* on the cosmos: this is what we learn from the Amerindian perspective. Consequently, tracking in a philosophically enriched sense must be *perspectivist*.

Perspectivism is an anthropological concept developed by Brazilian anthropologist Eduardo Viveiros de Castro, based on the symbolic system constituted by Native American shamanism. Perspectivism is an ontological attitude present in many peoples of the New World who share the idea that the world is 'composed of a multiplicity of points of view. Every existent is a center of intentionality apprehending other existents according to their respective characteristics and powers.'[3]

What, then, is really perspectivist about this philosophically enriched tracking?

During a tracking session in the spring of 2015, I was following trace by trace the trail of a wolf in the clay of a path, when I came across a stone slab, very long and wide. There were no scratch marks on the rock or the moss. So I looked up, and saw, in the distance behind the slab, a gap in the bush between some juniper trees which *might have* caught his eye and his desire to go straight on. By following this imaginary route, I quickly found his traces in the mud of this path; and there was this same claw, the front paw slit on the right: my female. I lost her again on the next slab, but the limestone

The discreet art of tracking

created gulleys which channelled movement and so, by choosing the one which went in the real direction of her movement, I found her trace again in one of these valleys. What happens on the ground and in your heart of hearts when you track a living being? Seeing through the eyes of another. A moment when species merge.

It sometimes seems to me that tracking, under our feet, in the forest, once our backs are turned, replaces one ontology with another: the naturalist schema becomes perspectivist, it crossbreeds with animism, mixes things up, creates chimeras. Just as magicians can whip away the tablecloth without moving the knives and forks on it, another map of living things on a 1:1 scale can be surreptitiously found under our feet, on the ground we are scrutinizing – another ontology to survey and share. Tracking is, on a small scale, a practice that enables us to circulate between worlds, between ontologies. How does this magic trick take place? What displacement takes place when we feel that we are seeing through the eyes of another?

It's quite tricky. Is it a transmigration of the soul? The human mind changing its body? This is all far too spectacular, too mystical, and too Western too. What is fascinating is that this question makes visible how the meaning of the concepts of mind and body changes radically depending on whether one is a naturalist or an animist.

Our cultural tradition has its own unique ways of thinking about these body-to-body journeys: reincarnation, metempsychosis, astral travel. Metempsychosis refers, for example, to the passage of a soul into another body, human or animal, plant or even mineral. The philosopher Apollonius of Tyana relates that, on seeing a lion, he recognized an incarnation of the pharaoh Amasis (according to Philostratos the Athenian, in his *Life of Apollonius of Tyana*, V, 42). The theme of astral travel is an expression of esotericism which refers to the feeling that the spirit is dissociated from the physical body

The discreet art of tracking

to live an autonomous existence and freely explore the surrounding space.

It's not at all this kind of experience that happens in tracking, it's much more down to earth. We're bending over some lump of excrement in the forest, and trying to find the trail we lost in the mud. There's not an atom of mysticism in this – except the mysticism of life itself. In the metamorphosis that we are trying to describe, there is also no possibility of freely exploring the surrounding space, like a ghost flying here and there: it's very constrained, and not at all a matter of flying over something or looking down on it from above. Constrained by what? That's the whole point. The fixed point of displacement that characterizes the experience of tracking is the body: you don't travel outside the body, and there's no one to travel outside of it. There is only body. But it's not the same body as that of the naturalists: body-matter conveys the disembodied spirit. So what is it?

Let's come back again to the enigmatic feeling that, when we track, we have *moved into* the animal: but what is moving into what?

In the same way that it's not the soul that changes body, it's not the perceptual apparatus of another animal that is borrowed by the human being, as seen in those virtual technology experiments with helmets or screens that let you see the same colours a canine eye sees, the types of contrast in a tetrachromatic bird's eye or a fly's eye. This is certainly interesting, but it always comes down to a spiritualistic and dualistic conception of the mind, which separates perception by the mind from action by the body.

In tracking, something else happens: it seems to me that what we see when we sense that we are seeing *through the eyes* of another animal is what its body itself sees, in the perspectivist sense, i.e. those 'affordances themselves' – the 'prompts' of its specific body.

Prompts are defined by the psychologist of visual perception,

James J. Gibson as the possibilities of singular actions of a specific body on a shared environment. The specificity of the body brings out particular types of prompts in the environment that surrounds us: each tree, stream, ford, field mouse hole, ledge or territorial marking of another creature *suggests* a different action depending on the form of life of the perceiver. A prompt is like an incentive to take a certain action, to behave in such and such a way, and the animal does not need to be conscious of it to follow it.

For example, that pass up there where the paths and the smells that rise from the two valleys converge is a prompt for the wolf: it prompts him to mark the territory so as to collect the landscape of smells which rise in clusters up to him.

Another example again, for an animal capable of gripping: a door handle constitutes a prompt to turn it, one that lies beyond other animals. For a territorial animal, an odoriferous marking constitutes a prompt to inspect and respond, while herbivores ignore it completely. A rare conifer in a Pyrenean beech thicket is a prompt for a bear looking for a scratching tree with perfect properties (and this is also how you can find its hairs in the heart of the forest, stuck in the sap of the conifer), while other animals do not even notice it. An overhanging rock is a prompt to geopolitical marking for a snow leopard, a prompt to signify its presence and its desire to its partners, but it is a prompt to an ibex, signalling a place to shelter from the storm, and a prompt for the Himalayan vulture to perch.

In the metamorphosis of the tracker, it is the prompts of the animal that we inherit; they are the ones that sometimes animate us when we have made ourselves sufficiently open, when we have worked hard enough on our own disappearance.

In the perspectivist attitude, we grasp the fact that the visible and the invisible relate to the visual capacities of the one who sees.

The discreet art of tracking

Indeed, strictly speaking, it is not a question of moving around in the mind of another, but in his *body*: it is his body with its own powers of seeing and doing which founds its perspective on the world; this is the great idea of perspectivism. However, this body is precisely an original effect of the eco-evolution which gives it its own powers and perspectives. For example, the Amazonian idea that 'the vulture sees the rotten meat that we find repellent in the same way that we see grilled fish' reveals that it is from its body as a scavenger, capable of metabolizing and neutralizing all pathogens, that the vulture sees and assesses what is desirable. It is in this sense that rotten meat is for him a prompt to tuck in and enjoy the feast, while for us it's a repulsive prompt to turn away.

This is another possible meaning of the expression used by Eduardo Viveiros de Castro, when he seeks to define the meaning of metamorphosis in animist traditions: it's not a question of slipping an animal physical appearance over a human mind, but, as we have seen, of activating in oneself the powers of a different body.

For a naturalist, the body is a vessel of flesh, a lump of raw, unintelligent matter, and it is the mind that bears the identity of the individual and therefore his perspective on the world. But here, what is interesting is that in tracking, it does not matter to us whether we move our intact human mind into an animal body-vessel: if our mind remains the same, as is the case in Western fictions which stage this kind of metamorphosis, there is no surge of intelligibility in finding the track, we don't see anything different, we don't see any more clearly. We are bounced along in another's body without seeing anything (and the problem is always practical: we have to find the trail, and follow it, not lose it).

It's not a transmigration of the soul, it's a metamorphosis. And a metamorphosis in the animist sense: it's not the physical body which changes around the irremovable mind as the bearer of identity (as in the naturalist conception of metamorphosis); rather, it's the perspective on the world that is metamorphosed, inasmuch as it is controlled by the body itself, but in a non-naturalistic sense: the body is *an attractor of specific prompts*. It's all about 'activating in oneself the powers of a different body', that is, accessing the prompts specific to another body. It's the body inherited from eco-evolution with its ethological finesse of feeling and reacting, and its own ecological relationships. In the metamorphosis of the tracker, it is the other's prompts that are being sought. When, as sometimes, it works, we see that col for a moment just as the wolf sees it, it prompts us as it does a wolf, it prompts the wolf that for a moment we have become. It's quite unspectacular, and practical in purpose, but it's no less intriguing.

All this is very difficult to formulate, we have to circle round it. To put it another way: it's not our mind that borrows the body of another animal, but our body that borrows their perspective, which is their body itself. And of course, in this metamorphosis we do not change our body in the sense of muscles, bones or fur (this is the myth of the werewolf in the naturalistic sense); we change our body but in the strictly animist sense: we adopt another perspective on the world with its own prompts.

Tracking comes down to borrowing, from time to time, and not at will, the body of another animal which is a perspective shaping the world around. It is because of, through, thanks to, on the basis of its body that an animal (including us) deploys a perspective on the world which surrounds it, that it isolates protrusions and prompts in the spaces it crosses.

It is in this discreet sense that we can pay lip service to the idea that philosophically enriched tracking is one of those practices that makes us change our metaphysics: a practice that quietly forces us

The discreet art of tracking

to become a little bit perspectivist, that is, say animist. The philosophical tracking described above by Georges Le Roy is indeed a naturalistic practice but one which, while we are squatting around an animal's print, behind our backs, transmutes the whole world around us: which tilts the naturalistic cosmology towards a more animist relationship with the living.

We can try to capture in a few formulas the specificity of this form of philosophically enriched tracking: perspectivist tracking is interested in living things in that it seeks, not first their Latin name, but the original powers of their bodies, their perspectives, the vital problems specific to them, the exorationality they use to resolve them, their founding ecopolitical relations with other living beings in their historicity and their plasticity, their potential mutualist relations with certain human activities and, finally, their mores, their uses, such as vehicular languages to communicate with them.

To qualify this particular way of being interested in living things, we could speak of neo-naturalism. 'Naturalism' means so many different things that the word is hard to use, but sometimes the ambiguities are useful. The two meanings that interest me here are Darwin's and Descola's. It's mainly a matter of being a naturalist in the sense that Darwin, during his round-the-world trip on *HMS Beagle*, observed living things and investigated all he saw as a fascinating mystery to be explored; this is the name of a practice shared by all those interested in nature as amateurs. In the nineteenth century and before, it was the name taken by the scientist, often an enlightened amateur, who took a keen interest in natural phenomena. The word probably appeared around 1527 to designate those interested in the relationships between living things, minerals and climate, but at that time it also described someone 'who follows his natural instincts'. When

The discreet art of tracking

the natural sciences became official and adopted more professional protocols, the word encountered the second meaning that interests me here: Descola's. 'Naturalism' then refers to a conception of the world where nature is essentially an inert matter deprived of interiority, explicable by exclusively physical causes and mathematical laws. Field naturalists are rarely naturalists in Descola's sense, but official research practices tend in certain cases to be limited to the logical classification of living things and their properties, of which national museums are the vectors and receptacles.

To specify the type of investigation of living things that I am trying to describe here, being a neo-naturalist is simply a matter of being a naturalist in the sense of Darwin's practices, *freed* from naturalism in Descola's sense.

The neo-naturalist is a field naturalist who practices his art without forgetting that he is an animal, without forgetting that those he investigates are much more than inert matter reduced to physicochemical causes alone. He does not seek first to pin the specimen down by its Latin name but to relocate it as a living cohabitant caught up in geopolitical relationships, types of modus vivendi, intimate weavings, with us and with others.

The contemporary neo-naturalist is not necessarily a scientist by trade; he is an enlightened and discreet amateur who can easily be tracked down online: he shares in obscure blogs his diplomatic knowledge about the alliances and enmities of the plants in his agroecological vegetable garden, his geopolitical knowledge of the living creatures who cohabit his garden at night, and which he tries to get to know better by setting delicate photographic traps . . . He shares his DIY way of not believing in naturalist ontology, of always postulating more in the living being than those who deprive it of everything, even if he does not know what its powers are and does not assert any

The discreet art of tracking

dogmas about what it is. He has freed himself from the confiscation of official knowledge on living things by accessing the endless bubbling knowledge of the Web, which is no longer compartmentalized in the minds of experts or distant libraries, but can be tracked down online with the same methodical ardour that one deploys to track down animals on their paths.

The originality of the neo-naturalist compared to his ancestors Wallace or Humboldt is that he knows that his art is an art woven from the animal powers that he finds and reactivates within himself, and that he finds outside of himself. It is with the panther's patience and his desire to find that he tracks the hidden behaviours of living things, that he strives to understand them. It is with the ancestrality of the selective gatherer found in the deer that he classifies and orders in his mind the different plants and their relationships. The classical naturalist is obsessed with the classification of living things; the neo-naturalist is interested in the cohabitation between living things. The gathering of clues here is no longer just a question of scientific objectification, of disinterested knowledge; it is a geopolitical practice. The question of knowledge is not for him that of disembodied truth, but that of the best cohabitation between living beings, in shared territories.

To be a neo-naturalist is to be a naturalist without naturalism, or, more precisely, beyond naturalism. The premise behind everything that is related in a neo-naturalistic inquiry, the specificity of the narrative, is that the narrator is an animal who is interested in other animals, a living being who is interested in living beings. Neo-naturalism is naturalism practised by a woven animal.

The discreet art of tracking

An art of seeing the invisible

The difficulty with all of this is that the characteristic of this universe of signs is that it is invisible or encrypted to the layman. There is nothing grand, impressive or explicit to see in it. And yet there is the collective joy of together finding a world of clues left by the animal, which reveals its habits, and its way of living in a habitat. More than an art of seeing, it is an art of imagining.

Those disdainful wolves of the Var, that inaccessible lynx that refuses to pass in front of the camera trap that we have installed with the greatest respect in front of the presumed entrance to its den in the Vuache massif in Savoie, trees which communicate with each other by invisible ethanols, corvids with their decisive intelligence that pass themselves off as simple-minded little sparrows, springtails camouflaged in the soils which they still manage to make so complex and alive: living things are sometimes tiresome in the way they make themselves invisible, without any explicit enigma, all their treasures hidden. Their silence, their art of disdaining the need to be interesting, requires so much energy if we are to make them talk without ventriloquizing through them – to restore their fascinating dimension and show how irreducible they are in their being – that they sometimes exhaust me.

It's so easy to think of them as *dumb beasts*, that is to say uninteresting: critters, machine animals, cases of mathematical laws, effects of mechanical causes. This is what our civilization has made of them. You have to fight to restore them to the vital enchantment that imbued them before language – but through language.

By exercising a different sensitivity to the living creatures all around.

We could call 'eco-sensitive tracking' any attention paid by a

The discreet art of tracking

living person to the signs of living creatures, to any indication of the intangible structures that govern them, to any trace that *concerns* them, involves their way of inhabiting and cohabiting, and calls for investigation.

We can read that the clear sky will soon be rainy from the low flight of the swifts that follow, like good insectivores, the descent of the aerial plankton drifting at the mercy of otherwise invisible barometric pressures. We can guess the presence of the goosefoot or the psilocybe fungus here, on these grazing soils, because they love the invisible nitrogen spread by domestic animals, or decipher the curiosity of the wolf from its pawprints which tell of its pause on a promontory. This is the kind of tracking I mean.

We can read from the remains of the red rosehip fruit that it has been a good meal for an invisible field thrush (which eats the flesh and leaves the seeds), or else that it has delighted an invisible European greenfinch (which does the opposite).

We can wonder, one knee in the hollow of the dead leaves, feeling connected to the humus, about the meaning of this clue: the red deer digs up tuber plants, it digs them up to eat *only a small part of them*.

You can recognize his two lower incisors which cut grooves in the bark, but he leaves the rest. He eats just a little of each tuber. Why? Is he managing his resources in his own way? Is it to let them grow back? Or is it because they are too tannic? It's a complete mystery.

Ecologically sensitive tracking does not mean you have to be in the wilderness: you can easily investigate the habits of the gulls that nest on the rooftops of Paris; your cat's nocturnal habits, and his impact as a predator who plays the innocent tomcat on the neighbourhood's biotic community; the complex ethos of the worms from the worm composter on the terrace, with which we maintain a delicate mutualist association; the friendships and enmities between plants in a

The discreet art of tracking

permaculture vegetable patch on the balcony. (Or you can investigate the ethological and ecological clues left by the humans buzzing in cities, insofar as they are, despite their stout denials, primarily living beings.)

My partner leaves a pot of rich earth on our terrace in the middle of the city, to make room for adventitious adventurers, these plants which use the wind, or the crops of birds, to explore new pioneer territories. It's a bit like those customs of nomadic hospitality where you leave a seat at the table empty, just in case. This year, one of those plant travellers, the common groundsel, has taken up residence with us.

A family of swifts lodges under the roof of the terrace. The parents fed the brood and then, following the immemorial call that binds them, left to spend the winter in Africa: as every year here, this takes place around 1 or 2 August. The chicks cannot fly yet. They will stay a few more days, a week, fly away without any help, perform a few exercises, and then head off to Africa on their own, following in the trail of their parents, in a migration they have never yet carried out, directed by mysterious senses. We sometimes feed them on the larvae of the flies that try to infest our earthworm allies in the worm composter. When their flight is low, we know it's going to rain soon, and we bring in the washing accompanied by their squawks. The groundsel is delighted. What blind man said we were alone in the universe?

By tracking, we must here understand any ability to access the intangible and the invisible structuring communities of life on the basis of empirical elements that are changeable and invisible to others. 'You cannot exist without leaving traces' is the tracker's magic formula. It flushes signs.

The discreet art of tracking

Tracking is probably a route to an ecosensitivity crippled by the 'extinction of experience': our loss of sensibility to and knowledge about living things. It imposes a formulation of the question that the eyes pose to the landscape, in the form of 'who lives here?', 'how do they configure this place as an entangled *being-at-home-ness*?'

Without the need to do metaphysics, tracking transforms the experience of nature, encoded by naturalistic modernity into a mirror of the soul, a bucolic place of healing, a backdrop for sports performances, or the background for a selfie, into an avid immersion of signs in shared and entangled habitats. Ecosensitivity involves experiences that consist in repopulating spaces emptied by the presences that constitute them, inhabit them, are linked to each other and to oneself.

Often in the morning, on the autoroute as I drive to work, I see a kestrel animated by its motionless flight, tracking, with its art of seeing what is invisible to me, its daily prey. This is a joy that remains intact each time: a gift of the kind already mentioned where nothing can be appropriated, a gift where no one loses the given, a gift without intention, but one which calls up in me an immemorial gratitude, that of the living cohabitants caught up in this community of destiny of the history of the living world. I feel joy at their beauty, strangeness, diversity, respectability. The mystery that their form of life constitutes reminds me of the mystery of mine. To exist as a living human being remains in my eyes an enigma, but this enigma is clearer, richer and more liveable in contact with other living beings, those enigmas.

Tracking also triggers in oneself a strange phenomenon which consists of an incomprehensible joy and a feeling of enlarged existence, simply because of the daily presence, pressing against us, of other living beings, cohabitants encountered by chance, sometimes annoying us, getting on with their lives following hidden motives of an irreducible strangeness.

The discreet art of tracking

The art of tracking comes down to flushing the invisible peoples who inhabit this world with us, by deciphering the visible clues.

Tracking as a mode of surveying shows up the unsuspected limits of our familiar hiking practices. In contrast to the form of attention developed by tracking, the hiker sometimes appears as a character insensitive to other living beings, a traveller who sees only himself while crossing the tangled habitats of others, which he has set up as his personal playground and source of spiritual healing. (As Montaigne said of a certain traveller: 'He saw nothing during his journey, for he had been carried away by himself.')[4] As if he were at home everywhere, in a world of things. When we track, we are often surprised by the strange habit of Westerners on walks, talking loudly, laughing without restraint: only at home do people allow themselves to be so noisy.

In tracking as an attitude, moreover, somewhat reluctantly, it seems to me that we see, feel and above all imagine things on a completely different scale from what we find in the aesthetic landscape of a postcard, or in the spectacular viewpoint. Things which give an additional dimension, in the mathematical sense, to the landscape, which hollow it out in all the ways in which it is inhabited by other forms of life, which are at home, which inhabit it with journeys, hunts, animal games, conflicts, displays, intimidation, dangers, fears, complex political relations, cooperations, alliances, types of modus vivendi and diplomatic pacts. Even if, most of the time, we don't understand much about them.

Attention to the animal landscape and to plant sociology, to the alliances of bacteria and roots, and the imagination of all these tangled lives, so strange and so intimately close, reveal another way of inhabiting nature, which becomes an unexplored diplomatic community.

Finally, what is intriguing about tracking is that it places us in

The discreet art of tracking

the same position as in the original forms of stalking and gathering, where we can only hope for an encounter, without forcing it: it is a practice that places us in a metaphysics of influence, without being able to apply our will to the encounter in order to provoke it. Tracking makes preparations for an encounter, but does not force it; it thus becomes an event of another magnitude. Tracking here restores an inner state that has become rare: the state of alertness, of a floating and loving attention towards the unforeseen.

At dawn, you can leave just to encounter, not knowing whom or what. It's a possible name for life.

An art of living together

If we separate, as we did above, tracking from the act of predation, it now becomes a certain form of attention. As such, one can then wonder about the nature of its fundamental uses for humans. These are not based primarily on hunting practices, although tracking largely originated in them. In a world where *Homo sapiens* has interacted for several hundred thousand years with a rich and ubiquitous fauna, tracking appears above all as a *geopolitical* practice. Much more than being a localized phase of predation, it is in my opinion a ubiquitous practice among the human inhabitants of ecological communities: a practice oriented towards the daily and primary question of cohabitation in a plural world. Its constitutive questions are: 'Who *inhabits* this place? And how does he live? How does he make a territory in this world? On what points does his action impact my life, and vice versa? What are our points of friction, our possible alliances and the rules of cohabitation to be invented in order for us to live in harmony?'

The discreet art of tracking

One may wonder what becomes of the fundamental political problem of sharing a common world with non-humans in this perspective. If the intangible use of the territory by living creatures involves habits, then the political problem becomes dealing with habits in intertwined and superimposed habitats. This raises the question of what a good habit is. It would be a co-habit, a co-evolution between the habits of several forms of life, which takes the form of an objective alliance, that is to say a mutualism.

It is this dimension of tracking that we are exploring here, as a contemporary practice of nature. The ecofragmentation which massively destroys animal habitats is not only an effect of major infrastructure projects; it is first and foremost the effect of our ignorance as to the invisible configurations by which animals and plants inhabit these spaces that we believed we could arrogate to ourselves.

This is one of the meanings of Michael Rosenzweig's projected 'reconciliation ecology':[5] to make the territories we inhabit habitable by other species, on a large scale, by making ourselves sensitive to their invisible demands, by seeing through their eyes. Habitable for them: that is to say, giving them space and time so that they can evolve (vary and be selected), and adapt to a world that has been massively transformed, and which in its broad outlines will never be the same as before.

The issue of tracking ultimately now resembles what it may have been during the long Pleistocene, an era governed by mutual vulnerability between humans and living things, a mutual vulnerability that has returned due to climate change and environmental metamorphoses. A world where coexistence with an abundant life required us to know *how* to cohabit with it, what habits could not be shed and what habits could be transformed, what powers we could compose with and what borders respect: a whole diplomatic geopolitics of the biotic community.

But there is nothing primitivist in this attitude: what requires diplomacy isn't the fact that we have returned to the Pleistocene; it is on the contrary the contemporary specificities of what some call the Anthropocene (and it doesn't matter what name we give it here): this new situation that is ours, this new form of cohabitation between different species on the surface of an Earth that has become crowded, intricate, connected. An era of mutual vulnerability where the wildest creatures live among us, impacted by the effects of our activities, by urbanization and climate change, by integrating as they can these transformations into their form of life, over which we have little control.

It is not about virgin or untouched nature deep in the forest, far from the cities. This is not a fully organized, artificialized nature put to work by industrialization and the capitalist economy. It is about something other than age-old nature: living territories deeply constituted and transformed by human activities, but where living beings have not lost their living power to *take back the upper hand*, that is to say to reconstruct new relationships with other species, with humans, new behaviours, new evolutionary directions. And this is true regardless of the legacy of destruction in our history. The new deal therefore amounts to tracking through the woven web of the Anthropocene, or, to use Anna Tsing's more joyfully apocalyptic formula, to tracking through 'the ruins of capitalism'.[6]

Several times we have followed the tracks of a pack of wolves that had settled right next to the Cadarache nuclear fusion reactor site in Provence. This centre, run by the CEA (the Alternative Energies and Atomic Energy Commission), at the confluence of the Verdon and the Durance, is the product of a complex geopolitical history: it

The discreet art of tracking

associates thirty-five countries, those of the European Union as well as India, Japan, China, Russia, South Korea, the United States and Switzerland. It was at the Geneva summit in November 1985 that Mikhail Gorbachev proposed an international alliance to build the new generation of tokamaks, those reactors meant to reproduce the sun's operation in a vacuum chamber. The project's budget recently increased to 19 billion euros, and the site houses a secret base. The wolves' tracks, meanwhile, ran along the barbed wire fences which secure the site patrolled by armed guards and, just behind these lupine tracks, the reactors, motionless mastodons which work to lock up whole suns in a box, loomed over the same landscape. Sometimes the passages of wild boars had pierced holes in the fences, and animals frequently rushed into this space forbidden to humans. Why had wolves, symbols of the wild, gone to live so close to this nuclear site? Was the forest quieter because tourists are reluctant to take their Sunday hikes in this kind of landscape? Or more full of game because hunters fear that the deer there have become slightly radioactive? It's a mystery, but the wolves were there, right up against the nuclear installation, in the interstices it induced. One summer, a camera trap even detected the fact that they had a vigorous litter of five cubs. We can see them playing on the dirt track that leads to the Cadarache facilities. So it looks as if you can also thrive in the shadow of a nuclear fusion reactor.

What we track in the contemporary world is never living creatures, out at a distance, in a natural world 'outside', but rather the interweaving of our stories and theirs, the biotic imbroglio that is the hidden name of ecosystems, as soon as we restore to them the contingent historicity, made up of unforeseen encounters, which constitutes them. There are not, there are no longer any intact living beings that move through a monotonous and repetitive 'nature', but forms of life

The discreet art of tracking

among us and simultaneously by themselves. The common buzzard (*Buteo buteo*) enters into strange mutualisms with motorways, feeding on the carcasses which they centrifuge out to their margins, and thereby playing a role in maintaining their fluid circulation. It is the environmental history of human activities and the responses of living things that can be seen in the behaviours we track. Postmodern tits offer new clues to read: they actively seek out cigarette butts to line their nests (nicotine is a powerful antiparasitic that protects eggs).[7] They invent new forms of life in the common imbroglio, influencing the history of transformation and decay which is our history. Snaking through these ecological imbroglios is the often indirect and far from sovereign vector of the tracker.

Recently, I spent several nights on the watch in the South of France observing the nightlife of a wolf pack as part of an action research project. Positioned in silence on a promontory in the middle of the plain, a thermal camera is aimed at the night, capturing the heat differential between the bodies in the landscape, and reproducing it in contrasts in the viewfinder. Then, lupine silhouettes made of harsh light appear in the clearings, play, repeat the rituals that comprise their existence, go hunting or patrolling their territory. But the uneasy aspect of this experience is that the camera in question is a military object prohibited for sale: war material, said to be 'sensitive'. It was designed for army border posts, and is intended to identify, among *other* things, migrants who would like to enter the territory illegally. The anecdote shows how this object materializes the porosity of modes of relationship: even if the goal is not the same, taking cameras designed to monitor migrants and using them to observe wolves makes one think. The technical device materializes what there is in common in our relationship to the forms of otherness that live right up against us. Moreover, it is in a military camp that we observe these animals: as the helicopters fly over us and the shells explode in the distance, we catch four wolf cubs playing in disused tanks. One night, the howls of the wolves were superimposed on the bursts of a machine gun. Between the tanks and the herds of sheep, all this human and technical fauna, the wolves settle down and take back control, in the minimal sense that they learn to live in and transform these environments which have inherited a long and complex past, which have inherited the ruins of a past in which living beings reweave themselves into new assemblages.

Tracking, in today's world, is not a bucolic experience: it's an ambiguous experience, woven from the acute awareness of the crises of cohabitation that blind economic activity induces, woven from our heritage of control over living things, woven from a desire to reactivate alliances in a changed world, to invent types of modus vivendi that are more liveable for all – woven also from the wager that living beings will be able to regain control in spite of everything.

5

Earthworm cosmology

We can go deep into the forests of Ontario or the Kyrgyz steppes to facilitate the learning of a philosophically enriched tracking, thus activating other ways of seeing by the force of our distant gaze – this is only ever just one potential detour on the way home.

It is among us that living beings prosper or decline, everything against us, and it is not necessary to go round exotic places or turn ourselves into Indians to track, that is to say to let ourselves be affected by the multiplicity woven from those ways of inhabiting which shape the common world. The form of attention, the precise quality of openness that practice teaches bring us straight home, in two enigmatic senses.

The first meaning, the most philosophical, is that the world that had been described to us, a silent and cold cosmos, violent and empty, governed by the absurd or the law of the jungle, and lit by the flickering glow of a civilizing human subject alone in the universe, a world which is not habitable: this world is only a construction of the mind. The world to which we are restored when we go tracking, almost unwittingly, is simultaneously more adventurous and more hospitable. Adventurous as it is repopulated by a thousand fascinating life forms with which we will have to start negotiating types of modus vivendi again. It is a world more demanding in diplomacy, but more welcoming by virtue of the very fact that its constitutive relations are not denied or torn, and that the cosmic loneliness of the moderns does not exist there – since we are after all surrounded by all the others, living things, rivers, bacteria, plants, animals, insects, oceans, wastelands and forests, which constitute us from within. Tracking, to this extent, is ultimately an art of returning home.

This point becomes intelligible, in contrast, if we question this other way of making ourselves at home, which consists of 'civilizing' a wild space. In colonial history, this *topos* is interesting: the colonist, coming from a different background, usually urban, feels that a new, wild territory becomes civilized only under very specific conditions. 'Civilized' means that the newcomers, ignorant of the ethology and ecology of the non-human cohabitants, can live there without the slightest vigilance and in all innocence (i.e. ignorance plus carelessness). A space is civilized if it poses no risks to me, even if I don't know it from within. And for that, we have to free it from all those strange beings, spiders, mosquitoes, beasts, bacteria, which become dangerous when I don't know how to live with them. In a sense, this is what it means for a settler to domesticate 'wild' nature or set up home in it. However, in the same territory, the natives are of course 'at home', and even more at home, without having to systematically destroy or control everything that overflows boundaries, or that makes a life for itself according to standards other than ours. It would be absurd to believe that these spaces, which have always been familiar to the natives as they roam around, are for them wild and inhospitable. Of course, both groups transform the environment to make it more habitable, but not on the same vector of 'civilization'.

Civilizing a wild place to make it habitable is probably the kind of concept that would appeal to an urban settler, because natives and rural people are at home there without distress: they have established forms of comfort whose specificity is that they require minimal vigilance, that is to say an attentive and diffuse openness to the unexpected, and a knowledge of the eco-ethology of the place's cohabitants. But the 'civilized person' wants to live in complete cosmic solitude, without having to be vigilant in his or her environment, now empty of presences, and without needing to know it, or negotiate with animal, plant, ecosystem and atmospheric powers, which he considers as *inferior*.

By a strange symmetry, such a person sees as a *liberation* from the constraints of the environment something which is probably, for the native who is the product of integrated multispecies backgrounds, an *alienation* in a dead world (remember the final scene in Kurosawa's film *Dersu Uzala*, where the old man of the woods, accustomed to arguing with the fire and negotiating with the beings of the forest, finds himself alone in a room of bare concrete, staring into the empty fireplace).

A story told by the Dena'ina Indians of the North Pacific advises the walker lost in the forest to call the wolf for help in finding his way. Compared with *Peter and the Wolf* and *Little Red Riding Hood*, this is a pretty pure reversal of the 'lost in the forest' mytheme: the great danger in the latter is salvation in the former. These are different ways of making home in a world populated by other powerful life forms.

The difference then comes down to this minimal vigilance, this latent openness connected to the flows of the outside, an openness cherished by those who inhabit and are familiar with living territories, and who view this intimacy with irreducible, sometimes disturbing alterities, is not a curse, but the tranquil name of one's own home – the name of real life as woven and present to the world.

Civilizing an environment, therefore, does not describe any attitude of being 'at home' in some place; it is believing that to be at home, one must live while being able to ignore, despise, subject and then dominate the forms of life and ecological conditions of the environment. To be at home, you need to have extracted the human home from biotic communities, and to have coded the living environment as a constraint from which you have to be free. You have to have stripped it of its status as a donor environment weaving interdependent entities, that is, its status as a tangle of connections that liberate.

But there are also other ways of making the environment a home in which it is precisely this vibratory vigilance to the plurality of forms of life, this interminable diplomacy with others (which are only dangerous if they are ignored), which is the desirable form of inhabiting it.

Tracking brings us back home, since the panther's patience, the tracker's narrow-eyed vigilance, his sense of the wealth of discreet things, now readily apply to the less spectacular living creatures who surround us in more urbanized worlds, for example the earthworms from the worm composter installed in the kitchen. The worm composter becomes a fascinating object.

It is a box with different floors, in which the combination of earthworms and microorganisms breaks down the organic matter in kitchen waste to turn it into fertilizer. The exotic mores of earthworms, their strange form of life, the new kinds of alliance that we forge with them in an apartment are now made prodigiously worthy of interest by the strange story being told here, the attempt to 'enforest' onself. Living with a worm composter becomes something more than a small ecological gesture. The earthworms are captured by the perspectivist logic of tracking, and we find ourselves forced to understand that this practical interaction with them requires our being just a tad animist: in the same way that, according to Amerindian cosmologies, the jaguar sees blood as millet beer, what we see as detritus is seen and experienced by earthworms as a *feast*. The worm composter forces us

Earthworm cosmology

to be a little perspectivist: we have to maintain it while respecting the bizarre demands of its inhabitants so that it won't smell of the rot that the earthworms prevent. Their strange life form consists, for example, in breathing *through their skins*, it is a perspectivist knowledge that needs to be acquired to forestall any mistakes: any oily liquid that we give them prevents them from breathing. Cohabitation comes at the cost of a delicate knowledge about their way of being alive, which must be tracked in our interactions with them.

Our kitchen scraps, in modern life, are thrown away as physical material that cannot be eaten – part of our Western habit of believing that all living energy retained in other living beings must come to us, and the uneaten remains are destined to rot. It's a practice which engages a metaphysics that we encountered previously during our meeting with the bear: that of the self-extraction of humans from the ecological community of the living, who set themselves at the top of the trophic pyramid, as if doomed to extract any resource and not to restore any of it to other forms of life. But the worm composter, beneath its humdrum exterior, is a metaphysical device, a kitchen-scale terraforming machine. It's a plastic bin with a cosmological scope, since it circulates living energy in the form of biomass: I am no longer the road to nowhere of biotic energy, the last and exclusive point of capture of the living matter that rises up to me. Now, some of it is returned to other living beings who can metabolize it, and from their prosperity is born the fertilizer and black gold (the earthworm tea enriched with all the nutrients of decomposed food) which will nourish the permaculture vegetable garden and its multiple biodiversity of insects and animals.

There is shamanism hidden in a worm composter. To get a glimpse of this, we need to take a brief detour through Siberian shamanism. If we follow the definition given by anthropologist Roberte Hamayon,

Earthworm cosmology

this shamanism is characterized by a cosmology of the circulation of the flesh which was briefly mentioned above. 'The system of exchange of which the hunter is the taker ensures the circulation of flesh between the different worlds: first the raw flesh of the game whose ritual reduces the murder to a catch of meat which nourishes human beings and, at the end of the circuit, the dead flesh of the remains of the hunter who returns to (super)nature.'[1]

In traditional shamanism, a whole series of practices thus ensure that the energy taken from the forest (in the form of hunted game) returns to the forest (in the form of voluntary death where the hunter offers his remains to scavengers, or in the symbolic form of disease and old age, which are thought of as a devouring of the flesh by the spirits of the forest). This helps ensure the renewal of the generations of humans and non-humans. These are practices, symbolic or effective, of *reciprocity*, which put their seal on another cosmology. Another way of relating to living things: a cosmology without human self-extraction, dedicated to perpetuating exchanges between living beings for the prosperity of each and every one.

Without adopting the hunting lifestyle which characterizes Siberian shamanism (which would be absurd), we can go so far here as to imagine cosmologies which no longer set us up as diodes for living energy, that is to say as receptors in which *energy goes only in one direction*, from the living world to us and never from us to the living – cosmologies that no longer set us up as transcendent living beings at the top of the food chain dedicated to monopolizing all the energy flows produced by ecological dynamics, and leaving to rot whatever we do not feed on. Yet, the worm composter, however comic, at first glance, this idea might seem, is a shamanic object. A thought experiment is enough to demonstrate this. Earthworms and the communities of bacteria associated with them feast on your hair, they feast on your nails. If this may be repugnant at first glance, it is because we have incorporated the metaphysical heritage that sets us up at the top of the food pyramid, as *uneatable eaters* – a philosophical construction that Val Plumwood has laid bare, and which is now visceral. Nails and hair, it is indeed our living matter, produced by our bodies thanks to the energy ingested, which we can restore to earthworms, and the latter restore it into fertilizer, which will go to nourish the communities of living beings of a vegetable patch or an agroecological garden, the fruits of which will nourish even passing birds in a circulation of the sun-flesh.

Through this alliance of living and active perspectives on the world, blocked circulations are freed and rewoven, even on microscopic scales.

As it is difficult to invent entire cosmologies out of nothing, we might as well start from accessible things: practices and their capacity to unfold, on their own, another world around them – slowly, gropingly, to bring out new cosmologies, habitable and finally liveable in, from the vermicompost bin.

In these strange forms, it is a cosmology of the circulation of living energy in biotic communities that is here reconstituted, against a cosmology of the appropriation of living things as a dead resource lying at hand, the residues of which are thrown away as mere matter when they are actually potential life for others, if the reciprocal loops are reconstituted, invented, imagined. This isn't a symbol (a symbol doesn't nourish your vegetable garden and its commensals). It's not a revolution either. It's a microscopic arrangement to help us move forward on another map, to change the base map of the world without any fanfare, to reconfigure our gaze and our actions, by contamination, with regard to everything that matters to us.

According to my partner, from this strange alliance with the worm

Earthworm cosmology

composter there sometimes emerges the strange sensation that the earthworms are watching over our common vitality. When we feel a bit guilty about not giving them much to eat for several days, this is a sign that we have fed ourselves on non-fresh produce, that we have neglected vegetable food. They eat little when we eat poorly. The loops of reciprocity, in multiple senses, are everywhere, like unexpected cycles of a shared causality.

There are other practices which produce similar effects, such as all those which consider relations with the living in diplomatic forms. Wild picking, in town or elsewhere, is one of them: when we practise it by learning to restore, for example by promoting the dispersal of the plants we have picked to imagine reciprocities, it reconstitutes something like a cosmology of the horizontal circulation of the flesh (solar energy converted into biomass by the great ecological processes).

In one Sioux song, the singer tells how he is transformed into a bear.[2] He sees himself advancing in the meadow, his foot transformed into a paw, and the song says:

My paw is sacred
There are simples everywhere
My paw
There are simples everywhere

My paw is sacred
Everything is sacred
My paw
Everything is sacred

Earthworm cosmology

Simples are the wild medicinal, aromatic and nutritious plants found everywhere in wastelands, forests, concrete cracks and meadows. Researchers have discovered in El Sidrón the remains of a young Neanderthal, who died some fifty thousand years ago; his dental plaque shows that he habitually chewed on poplar buds, whose analgesic and anti-inflammatory properties have recently been discovered by biologists. It is quite mysterious, when you think about it, that wild plants, which have evolved alongside us, conceal substances that relieve us, heal us, cheer us, make us live. Plants that we have neither cultivated nor selected, and which spontaneously, when we know how to find and use them, have for 300,000 years provided human beings with remedies ready to hand, offered and impregnable, in the intact mystery of their generous origin.[3] In a world where 'there are simples everywhere', it is not absurd to claim that 'everything is sacred'. It is indeed another cosmology which finds expression in this phenomenon, and in this song: the natural world is not initially an inhospitable savagery to be civilized by the sweat of one's brow, it is not an absurd cosmos of inert matter lying ready to hand; it is first and foremost a donor environment that eco-evolution has made surprisingly lavish for everyone. It is a home that nobody can appropriate, because home itself is not a habitat of inert physical matter: what we inhabit is the intimate weaving of other inhabitants.

These discreet cosmologies are reconstituted by practices nourished by our imaginations, endowed with a real power of erosion with regard to the world we have inherited and that we experience as dead. A power that is *at least* capable of operating shifts in our mode of life, and in the forms of commitment *to* what we hold dear, to what deserves to constitute us, and *against* what we know deep down will destroy us.

The diplomatic art is an art of weaving stories with practices, a firm intention to work with the world's harsh resistance, with all that spreads: good techniques, good ideas that don't belong to anyone, discrete methods for making experience more fluid, but methods that come laden with other cosmologies, in the way a fetish is laden with magic. It's a way of warding off all that is merely derisory by wagering that practices which can incorporate other ontological maps will have their own strength.

All these practices, and a thousand others, begin with an exacerbated sensitivity to other ways of being alive, and with a serene animism that takes the form of a working hypothesis without any mystical gobbledegook: other living beings are not things, they are centres of acting and suffering, they are points of view woven into and acting on the world according to their own standards, they are beings with two faces, in the minimal sense that they have something like an inner inside, they have interests – even if their insides and their interests are not thinkable on the same model as ours. To have an inside here simply means not to be a thing, but a centre that configures the world around into a proper living environment. And this centre of action that comprises them, these perspectives, this inside, are invisible. We are doomed not to see them, any more than we can see from within the feelings of the people we love. But nothing exists without leaving traces: this is the *arcanum* of the tracker, his secret law. The invisible leaves visible traces. The insides inaccessible to other living things, the wolves, bacteria and fungi that support the soil, trees that communicate, the bees that orient themselves to pollinate the world, manifest themselves through the visible clues they leave, through everything that takes place and that cannot be explained other than by the quiet postulate of this invisible life. It is these traces that call for tracking. Tracking is just the name we give to paying attention to the visible marks of the invisible insides of other life forms, of their way of being alive, be they earthworms, simples or panthers.

By practising tracking as an attitude of intense openness to other's art of sending signs, of being at home, among us but by themselves, it becomes possible to have another experience of the life all around, like an omnipresent community whose customs must be known if we are to live with it and from it. It creates a feeling of wholeness that I cannot explain. In contrast, one may wonder whether Pascal's 'diversions' are not strategies of avoidance meant not mainly to help us forget our own mortality, but to fill the absence of this life. 'The eternal silence of those infinite spaces frightens me', says Pascal,[4] thereby expressing the archetype of the feeling of cosmic loneliness of the moderns – but why does the ethnography of the first peoples who still live together in their ecological community *never* evoke this anguish of the silence of the world and our cosmic solitude, which is for Pascal the universal human condition?

Pascal saw the symptom, but his diagnosis is perhaps wrong: it is not our mortality, or the feeling of absurdity in a silent cosmos, from which we need to be 'diverted', but the silence of a living world transformed into a storehouse of mute things and resources ready to hand. The silence of the infinite universe is not a fundamental human condition, but a late effect of the loss of contact with our ecological communities, probably initiated in the Neolithic, then modified and intensified by modernity. The living universe is not silent, it is saturated with signs, it is always a complex song, subject to necessary interpretation. It is probably the urban character of Pascal's life, and the nascent naturalist ontology made possible by the way human beings warded off biotic communities and other points of view – spatial (settlement in cities), ontological (the refusal to recognize an ontological consistency in all that is not human), and technical (the generalized instrumentalization of the living by the moderns) – which make it possible to fantasize a silence, to create the deafening silence of infinite spaces. Because behind the fence, or in the vacant lot, or at the window, for whoever wants or knows how to listen, there is no silence.

One could argue that our feeling of cosmic loneliness is proportional to our blindness in the forests: if the forest and, beyond that, all the places where the living world finds expression (that is to say, everywhere – outside of shopping centres) could become the main sites of the crisis of metaphysical loneliness (as sung by the Romantic poets), it is because we were blind when we surveyed them. It is perhaps because we no longer really know how to recognize and read the traces and signs of life that we sometimes have the feeling of being deeply alone in a forest, and indeed in the world. In other words, it is not the living universe that is silent: it is we who can no longer hear or read it. The living world is no longer a network of meanings in which humans can orient themselves, as if they were at home. It is a still life, an aesthetic landscape devoid of its ecological, ethological and evolutionary meanings and mysteries, and thus a blank page on which to inscribe our metaphysical anxieties.

Lactuca serriola is a fascinating wild plant: the layout of its leaves indicates the north/south axis. It is a living compass which indicates the cardinal points to the lost traveller who can recognize it. This is because this plant orientates her leaves in such a way as to prevent them from being burnt by the sun. For those who seek it in an urban wasteland, for those who decipher, on the side of a road, the strange symbiosis between a common buzzard and a motorway (I clean you, you feed me), or who thank the deer for the path that leads them to safety in thick copses, and dig it deeper for them, or can compose successful alliances of food plants on their balcony, there is probably no cosmic solitude.

We change metaphysics only by changing practices. Consequently, the practice of tracking, among others, acquires another dimension: that of reclaiming a delicate place in the living world; of reading, giving and exchanging signs; of understanding and dealing with the creative richness of living things in their relationships with humans; and finally, of piecing together a cosmology a little more lovable than ours, a rather liberating feeling, perhaps — that of living as a living creature, rich in its inner heritage, woven into other living beings.

6

The origin of investigation

So far we have been telling a story about a practice that has been pieced together, one that is unstable but which has been developed to transform relationships with the living world that were ultimately found to be too poor. This has been at the level of individual practices. If we change the scale radically, we can question the role of tracking in the process of hominization which has lasted for several million years and which has made us the humans that we are. We can investigate tracking on a geological time scale, and its place in the emergence of the strange species that we are. The anthropologist Louis Liebenberg formulated the hypothesis already mentioned above that tracking probably played a major role in the emergence of human thought, in the precise form of investigation. He suggests that some of the most unique intellectual abilities of humans may be effects of the pressures of selection on the scale of evolutionary time, pressures that have focused on the activity of tracking prey.

To tell this story, we must remember that, as a species, *Homo sapiens* is the complex product of coevolutions with the various species it has encountered in its evolutionary trajectory, right across different ecosystems. Every species is the history of a singular drama. There is an absolute singularity to the trajectory that constitutes the species of the dolphin, like the termite or the wolf. What is the oddity of this constitutive trajectory of *Homo sapiens* that makes it such an intriguing species for us? The central idea here is that

The origin of investigation

one of the decisive phenomena of hominization lies in the transformation over two million years ago of a certain primate with a predominantly frugivorous diet, in an African forest ecosystem, who changed into a predominantly carnivorous omnivore in a savannah ecosystem. This historical combination – a frugivorous primate who *turned* carnivorous – is a singularity of the human species. It may help to understand some of our quirks more acutely than by drawing zoological comparisons with other monkeys, which place too much emphasis on phylogenetic proximity with our only primate cousins.

The evolution of intelligence

In his field studies concerning the tracking practices of Bushmen hunter-gatherers of the Kalahari, Louis Liebenberg develops a hypothesis on the role of tracking in the emergence of certain human cognitive skills that I would like here to reformulate, modify and enrich with new aspects.

A wide range of data suggests that the genus *Homo* actively hunted, at least since *Homo erectus* appeared about 1.9 million years ago. Numerous tool marks left by meat-cutting operations on parts of the animal that would have been eaten by other animals before our ancestors could gain access to them if the prey had simply been scavenged, leave little doubt as to the existence of active hunting. But how was it done?

When we imagine our pre-human ancestors, we accept without much thought that they could have caught big game with stones or hardened spears. However, it turns out that, given the athletic and defensive skills of large ungulates, this hypothesis is unlikely. The bow and arrows appeared much later: probably after the emergence of *Sapiens* (the oldest bows so far found date back 64,000 and 71,000 years). A spear can only be thrown effectively from ten metres. Before the invention of the spear thruster and the bow, it is unlikely that *Homo*, given the vigilance of his prey, could have come close enough to large ungulates to kill them.

This is what leads us to postulate that the hunting technique that has been ubiquitous for the longest part of the history of the genus *Homo* is persistence hunting. This is a technique that can still be observed today among some hunter-gatherer peoples, notably the Koi Bushmen of the Kalahari. It consists in looking for the fresh trail of an ungulate, and then following it, chasing the animal which moves systematically when it smells or hears the tracker on its trail; this can go on for several hours, until the animal is immobilized by its own hyperthermia (i.e. the heat produced by the activity of its muscles). It is then at the mercy of the hunter. Indeed, the large ungulates of the savannah have means of regulating body heat induced by the effort which are less effective in the long term than those of humans (like the big cats, they have more effective thermoregulation for sprinting). It is therefore by inducing in the animal a rise in temperature such that it can no longer flee that the hunter can approach it closely enough to kill it. The hunt often lasts eight hours, up to twelve hours in rare cases. The animal is then finished off, at close range, with a spear blow to the heart.[1]

This hypothesis, which holds that persistence hunting is a very old and enduring practice, is reinforced by one of the most spectacular phenotypic peculiarities of the genus *Homo*: the progressive loss of the fur, which makes it a 'naked ape'. However, the latter can be interpreted as an adaptation to the thermoregulation by sweating required by the persistence running specific to this kind of hunting: bare skin allows heat to be evacuated more effectively than the fur of the prey being tracked and the pursuer can still be fresh after hours of chasing while the pursued is overwhelmed by its own body heat.

Persistence racing still seems to be visible in the *Sapiens* phenotype: biomechanical adaptations favouring balance and speed, and optimizing the movement of the pendulum, are signs of this. When the primate ancestor of *Sapiens* began its orientation towards a massively meaty diet, it was thus subjected to pressures of selection that favoured its aptitudes for rapid and enduring movement, shaping the singular profile of the walker and bipedal runner with bare skin. If the effects of these pressures of selection, present for hundreds of thousands of years, are still visible in our bodies, we can assume that they are also visible in our minds, and define us as much as bare skin.

Because indeed, in order to succeed in this type of hunting, it is certainly necessary to be able to run for a long time; but above all to run in the right direction. The hunted animal is not in the hunter's line of vision: all the latter sees are its tracks. So that, together with bare skin and a runner's body, natural selection also had to apply to the ability not to lose track of the prey.

The argument we will follow therefore consists in isolating a correlation between a certain technique for obtaining food – persistence hunting – the cognitive skills it requires, and their role in the process of hominization. How are the intellectual skills specific to long-distance tracking (systematic tracking and speculative tracking, as defined by Liebenberg), as selected by evolution in our ancestors, the beginnings of our art of thinking?

Seeing what is no longer there

I would like to start with a reflection on what is still happening upstream of Liebenberg's reflections, in the evolutionary event that

The origin of investigation

led a certain fruit-eating primate towards the food-capturing act of tracking.

There is no need to compare our intelligence exclusively to that of other primates, because they are not hunters or trackers, as *Homo erectus* (or *ergaster*) must have been for some two million years. The whole problem then comes down to the fact that we are the bodies of fruit eaters who have become carnivorous trackers, that is to say visual creatures condemned to find invisible things. The eco-evolutionary combination that underpins our cognitive identity as living thinkers is the encounter, in a specific form of life, of a frugivorous social primate past (a weak sense of smell, a strong eye, and a cunning theory of mind when it comes to determining the intentions of others) and new ecological conditions involving new pressures of selection: bipedal life in the savannah involves an omnivorous diet with a carnivorous tendency that *demands* tracking. This is a key to the mental powers of the human being, that combinatorial animal.

To find invisible things without smell, the solution was to awaken the eye that can see the invisible, the mind's eye. Our coevolution with our new prey would have induced the establishment of original cognitive skills. So it is not the change to a carnivorous diet that is the central event (although it does play a role in the protein supply probably needed to fuel a large brain); nor is it hunting as predation and devouring (although it does play a phenotypic and ecological role), but *tracking*.

How do other predators access their prey? Raptors are predators in which sight dominates, but their hunting problems are very different from those of the walkers that we are: the clearing offered to the gaze by aerial life implies that the quasi-optimal evolutionary solution comes down to selecting keenness of sight, for the zenith perspective makes it possible to see the prey from a great distance, and to track it *by eye*.

Let us refer to the ecological living conditions of the terrestrial carnivore (the wolf or the panther, for example). The vast majority of these animals are originally endowed with a powerful and discriminating sense of smell. The wolf roams its territory without prior knowledge of the location of its game, and when it comes across a 'trail', it follows it until it finds its prey (it is said to 'set foot' on the animal being followed). Its ability to reach its prey by following its track is the direct result of pressures of selection. The track is made up of visual and scent stimuli. For the hunter by smell, the stimuli of smell alone trigger animal identification by neural ignition. Also, for the hunter it is smell alone that allows him to know which way his prey is going: it is the sense where the intensity of the stimulus is strongest. The vernacular experiment of tracking down a bakery by following the odour of hot bread on a street illustrates how tracking by smell requires few abstract cognitive operations to orient itself and determine its likelihood of success. When the Eurasian lynx tracks the snowshoe hare in the snow and the trail goes cold and stale, he goes off in the wrong direction fifty percent of the time: he can hardly read the direction of the track with his eyes when odoriferous signs are absent.

Now we have become terrestrial carnivores, but without a strong sense of smell (fruits and leaves don't need it, because they do not flee); our sight is powerful, but we are tied to the ground: the sight of a raptor would not be enough to pierce the canopy of the woods or the roundness of the Earth. The hypothesis is as follows: the singular vital problem at the origin of *Homo*'s cognitive specificity, a problem for which selection has gradually found ever more refined adaptive solutions, is: how can you track food when you do not have the sensory adaptations for this task? In us, because we are this fruit-eater turned tracker, visual performance must be supplemented by that of the inner eye.

Humans are fundamentally visual predators; and it is the relationship between his own form of vision to his ungulate prey that governs his relationship to tracking, that is, to obtaining food for much of his history. Because his eyesight is penetrating enough to follow a trail, but not sharp or panoramic enough to hunt only by sight.

It is this human condition of having good eyes but eyes that are fixed on the ground, rather than having flying eyes or a keen sense of smell, that creates the platform for the forms of tracking that are probably the source of some of the aptitudes of human thought. A weak nose, eyes fixed to the ground, slow running speed: to reach the animal, you have to follow it over long distances without seeing it; that is, it has to be tracked. It is the power and limitation of human vision that awakens the eye of the mind, the most powerful in its radiation of cognitive effects that living beings have so far developed. By comparison, the wolf does not have the same sensorimotor scheme. His sense of smell is very highly developed. What does the wolf *see* when he sees an ungulate pawprint? This is a very difficult question. Opinions differ. Is he able to decode its origin? That is to say to mentally evoke the image of the prey whose trace is the present-absent symbol? The question, I believe, is rather: Does he need it? The hooves of an ungulate secrete odours, and the wolf sees the ungulate more clearly in that odour than in the mental image inflamed by neural ignition at the *sight* of the pawprint, like a human. Indeed, a powerful smell has a similar evocative effect on us, even though our sense of smell is so derisory. Let us therefore postulate that the wolf is happier following a trail of scents, not because his eyes cannot, but because the other trail is more alive. It is because the odour tracks are

The origin of investigation

not very alive for him, being poor in evocative power, that the human has had to enhance the symbolic power of *evocation* of the dead trace visible in the sand, and this alone, thereby enhancing the assemblies of neurons which create the mental image in an abstract way: this marks the birth of something like the proto-symbol, set up by the activity of interpretation which *reads* in it more than it sees.

Because the point of the problem is that if we can be satisfied with recognizing an animal by its smell and following it by going to the place where this smell is strongest, the trace in itself evokes nothing: it must be deciphered, interpreted, read. The side effect of not having a good nose is that the eye is doomed to do more work, plugged into the brain: it first has to determine, for example, which way the animal has gone, on the basis of the asymmetry of the traces. This is already an elaborate intellectual act that is demanded simply because the image is less lavish in direct information than is smell.

Imagine a trace in the sand: for the visual hunter, the trace needs to be read, to be translated, that is to say to be interpreted as a sign. He is forced to look for the pawprint, which familiarizes him with the phenomenon of the sign: a present element which beckons towards an absent one. These aptitudes constitute some of the early stages of symbolic faculties: a need to learn to read traces, that is to say to interpret. Anyone who has had the experience of looking at ungulate tracks in the forest remembers having to learn to decode the direction of the animal from the asymmetric shape of the track. The amount of information contained in this ideogram can be staggering: species, age, sex, direction, state of health, individual identity, emotional state, current activity.

In tracking, we see the potential emergence of decisive cognitive skills, which revolve around the power to see the invisible, for example the destination of the animal or a sequence from its past. A

The origin of investigation

straight path, for example, indicates that the animal is heading to a specific location. When the animal returns to its den, its trail is thus often quite straight; if the trail meets with other, older tracks of the same animal, we are getting closer to the den. The animal wanders round foraging on the outward journey, but he knows very well where he is going on his return. A good tracker can read the trail of a predator who is hunting and then resting in the heat of the day, taking a bit of meat, then heading off again.

> From a certain track, the stalker can say: the lion here was resting, he heard a female's cry, got up and trotted up the dune to hear better, waited, then went back to find the female. Given this hypothesis, the tracker goes to look for the track, and two hundred paces away he comes across the tracks of a female, whose urine leaves show that she is in heat. There, the tracker finds traces of another male, and a fight between the two males. Then one of the males runs away, and the other leaves with the female. All that was visible there were the traces of a lion rising and trotting towards the dune, pawprints in the sand. But their shape and gait showed that he was not hunting, for his trotting was quiet and not close as when he lies on his hind legs to be discreet. If he goes up the dune and stops, it is to listen. Then the way he moves shows that he is attracted to a female.[2]

Tracking thus consists in reconstructing and extrapolating a history of animal activity that is much richer than what traces alone show, for it provides access to the invisible. The tracker therefore sees the invisible, in the literal sense of the word, like the doctor who diagnoses the presence of an invisible bacterium hidden in the viscera simply by deciphering a bundle of symptoms on the surface of the body. He accesses the invisible through the discreet visible traces that the invisible leaves (and nothing exists without leaving traces).

The origin of investigation

Practical tracking manuals are strict on this point: identifications made at first glance are often wrong;[3] a good tracker must place the sign he finds in his heart of hearts and arrange it in a series, in a critical constellation with other signs, before determining the identity of the prey. Tracking therefore requires systematic investigation and the systematic suspension of judgement before obtaining enough signs to confirm.

The tracker 'explains that one should not look too hastily at the tracks, because that would lead one to see them as other than what they are. He asserts that one must study the traces carefully and think carefully before making a decision.'[4]

In fact, when we're out in the field, leaning over a track in the clay or the snow, we often simply can't read it or decipher it. Several times we are wrong and the rest of the track proves as much to us. We are forced to learn to accept uncertainty, to remain in doubt. We learn to resist the urge to conclude to escape the discomfort of not knowing. It's hard to learn to say 'I don't know' (when this is often the smartest thing to say when faced with a trail). However, it is ultimately liberating, and especially when we learn, through the investigation of anthropologist Nastassja Martin for example, that Amerindian trackers of the Far North, the Gwich'in, have the serene habit of saying of a pawprint, 'It's a wolf. Or something else. Or not. Maybe.'[5]

Trackers' remarks often refer repeatedly to the need to reflect, to take your time before deciding to identify the animal. There's a systematic suspension of judgement which is necessary in the field, and which pertains to this exercise. We can therefore ask ourselves: If tracking was a foundational activity of human cognitive forms, and if suspense is necessary, what might this have induced in daily cognitive tasks?

The origin of investigation

One could hypothesize that this is one of the origins of the suspension of judgement as a mark of sagacity. The relationship between visible and invisible, which is the human cognitive problem, demands that judgement be suspended. Presence-absence is the problem of stalking. It is also the problem of attributing intentions to other congeners – determining an invisible intention from visible behaviour, reconstructing a past action from pawprints. It's as if these two ecological living conditions (as tracking animal and as social animal) had induced selection pressure on the ability to mentally reconstitute the invisible links between visible fragments taken as clues. It's a specific cognitive problem that probably helped shape human beings.

There are two types of tracking.[6] *Systematic tracking*, which consists of following the animal trace by trace. This tracking nevertheless requires knowing how to read, interpret traces and suspend judgement. Systematic tracking is sufficient over short distances, but as we have seen, *Homo* has become a runner, he practises persistence hunting. Over long distances there always comes a time when the animal is lost from sight (rocky ground, river, the crossroads of animal routes, etc.): this is where the need for speculative tracking comes in. From this point of view, evolution has not only established our systematic tracking capabilities, but also those specific to *speculative tracking*:

> In order to reconstruct the animal's activities, the emphasis is primarily on gathering empirical evidence in the form of spoor and other signs. Speculative tracking involves the creation of a working hypothesis on the basis of initial interpretation of signs, a knowledge of animal behaviour and a knowledge of the terrain. With a hypothetical reconstruction of the animal's activities in mind, trackers then look for signs where they expect

The origin of investigation

to find them. The emphasis is primarily on speculation, looking for signs only to confirm or refute their expectations. When their expectations are confirmed, their hypothetical reconstructions are reinforced. When their expectations prove to be incorrect, they must revise their working hypotheses and investigate other alternatives.[7]

The tracker formulates possibilities, which are enriched by positive and negative feedback. We can here see the serious sense in which Liebenberg sees himself justified in thinking that tracking was something like an origin of science. But the idea of science in general is still too vague. Although he does not make the distinction clearly, it is obviously not a question of science as a mechanism of historically dated knowledge, but of something else: specific cognitive skills required by a particular form of inquiry, called 'rational', and their methodical articulation.

The hypothesis that I am making here, connected to Liebenberg's, is the following: the human being appears and develops cognitively from the point of view of these aptitudes, because humans fit into an ecological niche in which foraging requires *speculation*, taking the word in a precise sense. It demands the process of inquiry as it articulates the three fundamental inferences of human logic: abduction (hypothesis-making), deduction and induction.

The first skill specific to speculative tracking is indeed the formulation of hypotheses and their testing. Speculative tracking is about creating a working hypothesis about something inaccessible to the senses: the invisible. This is abduction. Then it is about *deducing* what one ought to observe in the visible, in the empirical realm, *if ever* the hypothesis were true. Finally, it is about actually seeking it out in the field to test the hypothesis, repeatedly, in order to be able to generalize the knowledge involved. Now this articulation of the

The origin of investigation

three fundamental inferences of human logic (abduction, deduction, induction) coincides precisely, in this order, with what the pragmatist logician Charles Sanders Peirce calls the 'scientific method', or method of inquiry.[8] Persistence hunting is thus the form of foraging of a mammal that involves the cognitive skills that gave rise to investigation. Investigation is here to be understood in the pragmatist sense of the process of seeking reliable beliefs which articulates in a precise order the three inferences of human logic.

The role of thousand-year-old pressures of selection specific to tracking when these pressures are brought to bear on the emergence or orientation of certain human cognitive skills, of the order of logic, deserve to be systematically examined. Epistemologist Ian Hacking hypothesizes that while styles of scientific reasoning are historically dated, logical skills are prehistoric.[9] For example, absurd reasoning seems to play a decisive role in tracking:

> It is also important for trackers to know how to discern when there are no traces at all. On hard ground, trackers need to be able to tell if the animal would have left any tracks if it had actually been there. This is important because patrollers need to know when they are off a track. Let's say that an animal could have followed two different paths. If the tracker sees that there is no trace where there should be, then the animal has probably chosen the other route.[10]

The ecological conditions specific to the tracking of a hunter primate could have been a pedagogy for certain inferences such as reasoning by absurdity, a particularly elegant cognitive ability of the primate that we are. The presence of externalized traces as a support for abstract reasoning, such as the daily familiarization of a certain primate with logical problems of this order (here there should be traces because the mud is soft, but there are none, therefore . . .) have played a role as facilitators of the acquisition of these forms of reasoning, because it is probably easier to learn to reason ad absurdum with externalized visual supports (pawprints) than with abstract logical propositions.

With Liebenberg, we are witnessing an evolutionary history of the emergence of certain aspects of human thought: it suffices that there has been a selection pressure on the matrix of cognitive faculties required by tracking. These faculties would be the matrix from which, by exaptation, that abstract and living form of thinking known as human investigation took shape.

From empathy to imagination

In speculative tracking, once the tracker has grasped the general direction of the track, and knows that along this axis there is an animal road, a river, a key point, he will leave the track to go directly towards this key point to find the way. To predict movement, you need to know the animal so well that you identify with him. You have to visualize how he moves from his own point of view.

This knowledge of animals is not just an inductive knowledge of their habits, their behaviours, their ecology; it is also an aptitude to transpose oneself into them in order to make hypotheses. From this point of view, we see a crossbreeding of the cognitive aptitudes specific to tracking with our own aptitudes as social primates to master a theory of mind, that is to say an aptitude to postulate intentions in our fellow humans, beliefs and desires, and to decode them. The theory of mind shifts (exapts) in the primate turned hunter, being applied not just to his congeners but to his prey. We can hypothesize that *Homo* finds his cognitive singularity in being a social interpreter oriented

The origin of investigation

towards tracking; in other words, he will use his selected gifts as a primate psychologist to interpret living beings *other* than his congeners. Tracking catalyses the psychological and social interpretive faculties, as it exapts them towards diplomatic activities with other living beings, in the sense here of understanding their way of life and their modes of communication.

In tracking, there is a strange alliance between empirical rigour and imaginary projection, storytelling and accuracy. You have to decipher clues and draw the right deductions in order to fabulate well, and then you have to fabulate well to orient yourself well. There is no opposition between inner reflection and an intense openness to the world outside: to track is to think in a concentrated way, but the whole process is outside of yourself, it involves reflecting on the outside, being fully stretched across the scale of the landscape you are walking through. With the tracking described here, there is therefore something invisible, there are mysteries, but without a world 'behind' the appearances, without any transcendence. The world is more than appearances, but it does not need essences or supernatural entities – there is more than enough meaning, riches, enigmas and beauty in the old appearances.

Speculative tracking involves following an imaginary route, sparing yourself the effort of examining each pawprint, visualizing the animal's journey through the bush through his own eyes. The expert has his eyes pointed towards the horizon; he does not look at the ground, he dreams it. That is, he looks for signs on the ground only where he has *projected* them to be. 'What would I do if I were you, animal?' (but if I were you in depth, with your desires and aversions, your prompts, your rhythm and your world): this is the guiding question he examines as soon as he gets lost, to find his bearings again.

The origin of investigation

The connection within a certain primate of speculative tracking skills and theory of mind implies that the prey-seeking activity coincides with what has been described above as a shamanic phenomenon: a form of displacement of one's mind into the animal's body. It is from the daily act of tracking, under pressure of selection, that human beings' original therianthropy seems to stem (therianthropia are half-human, half-animal figures, like the Egyptian gods): their immemorial power of interbreeding with the rest of the animal cosmos – through the ability to become the wolf that hunts and the antelope that chooses its path. For Liebenberg, this decentring from one's own body, a decentring required when involved in persistence hunting, helps to explain the adaptive value of empathy. He bases this intuition on the explanations given by Nate, the Bushman tracker who is one of his interlocutors, about the inner necessities of the hunt:

> Nate explains to me that the tracker must continually measure his or her own physical condition against that of the kudu – looking at the kudu's tracks, the breadth of its stride, the way it moves the sand indicates how tired it is. You have to compare the state of your own body to that of the kudu [. . .]. It is your personal feelings that tell you about yourself and the kudu – not paying enough attention to these feelings can lead to hyperthermia. This example shows the importance of empathy in the success of the hunt and thus its adaptive value in terms of natural selection.[11]

This metamorphosis into its most empathetic form is not a romantic reverie: it is, on this hypothesis, an aptitude formed under pressure of selection in the evolution of the genus *Homo*. We can see here in what sense animal diplomacy, as an attempt to access exo-rationalities hidden inside other animals, and more broadly to grasp the internal logic of living and non-living beings (oceans, mountains, skies), is

The origin of investigation

based on immemorial skills, which have partly created human beings in their cognitive singularity.

We are diplomats towards living beings, destined by our animal powers to understand how *everything behaves*, open to the perplexity of determining what to do with these powers. Properly understood, these do not imply any human exceptionalism. These are not signs of election, but animal oddities among all the rest. They do not enable us to look down from above on other living beings, but immerse us irremediably among them, right in the middle: they are relational powers.

An exaptation of tracking skills

Tracking, as a daily habit and a vital necessity, that is, one based on cognitive skills selected over several hundred millennia, has set in place the cognitive skills that are the 'exaptive reserve' of the faculties of *human thought*. By exaptive reserve, I mean that current aptitudes are probably a twist given to the uses of mental organs selected in evolution. The set of complex and heterogeneous pressures of selection that operated on human cognitive skills during hominization formed a reserve of traits that were not meant to do what we do today with our minds, but that make it possible – from mathematics to art and philosophy.[12] In evolutionary biology, two types of traits can constitute exaptations: first, traits selected for a first function, made available for a second by an unforeseen change; but also – and this is where the interest lies – architectural constraints induced by the emergence of a selected trait, which are available to help with acquiring a new function or a new use. The more complex an organ is, the more collateral architectural constraints it manifests. The brain is a singularly complex organ. As an organ with skills selected by

The origin of investigation

evolution, it is constantly available for a redefinition of its uses by those who use it. Gould traces this intuition back to Darwin:

> Darwin, who was not strictly speaking an adaptationist, recognized that the brain, although unmistakably constructed by selection for a certain complex set of functions, can as a result of its intricate structure function along an unlimited number of pathways relatively unrelated to the selective pressure that constructed it. Several of these pathways may become important if not indispensable for future survival in later social contexts (such as five o'clock tea for Wallace's contemporaries). Most of the things the brain does to foster our survival relate to the domain of exaptation.[13]

We must briefly distinguish, to be exact, between the exaptation of function and the exaptation of use. The first describes the moment when a trait acquires a new function in the sense that it is taken up in a new process of natural selection: for example, the feathers of birds, which find their origin in thermoregulation, have seen their function diverted towards flight; they were subjected to new pressures of selection specific to this function. The second, which concerns us much more here, describes the moment when a trait is diverted to a new use by an individual without being subjected to natural selection: for example, when tracking skills are exapted and used by an art historian to decipher Rembrandt's paintings.

Tracking has certainly induced pressures of selection that lie behind cognitive skills, some of which are still used almost identically today, while other of its architectural constraints have been exapted to unprecedented uses. Writing literature, exploring the infinitely small, retracing the history of the lost civilizations of Sumer – all of these uses of the human mind involve skills probably stemming in part from

The origin of investigation

tracking, diverted to new uses, which continue to be invented. Much of the reading of signs then seems to be derived from this natural habit of reading pawprints as signs of the invisible in the visible.

It is the spectrality of human intelligence, the ability to frequent from his bed, with his eyes closed, all the inaccessible, vanished, immaterial, distant corners of reality that makes the human animal so bizarre. He has an ability to formulate and solve past, future, theoretical and practical problems, haunting absent landscapes like a spectre, landscapes that he has reconstituted within himself.

Fundamentally, the present life form of a species is a subversion of old habits forced to grapple with new vital problems transmuted by its long historical development – a way of piecing an existence together. But the basic material is essential and here this material comes down to the cognitive skills of tracking as a mode of reading signs and reconstructing absence within oneself. Interpreting and reconstructing are two types of activity that are found everywhere in tracking. They predate the existence of written texts, which appeared only three or four thousand years ago. But writing becomes something much less enigmatic in light of this form of life known as tracking: it is a way of reading signs. The symbol becomes much less incomprehensible, because the pawprint as a clue is an intermediary between the association of iconic ideas (smoke and fire) and symbolic reference (word and thing); it is this which makes the transition. It is tracking that allows us to think of the conditions of possibility for the appearance of symbolic thought, of the spoken word, of the written word – all of which are avatars of the pawprint.

With the art of the original tracker, the figure of humanity's first investigator, we are probably witnessing something like the emergence of our intelligence. But it is an intelligence of great ecosensitivity: sensitive to the vibrational finesse of the living world, to its

The origin of investigation

variegated cosmos of meanings and interactions. It is an ecological intelligence that we have, as it were, left behind somewhere, once we started to think of the donor environment as Nature, then nature as matter, and locked ourselves into our human prison, losing contact with the great vital politics of the animal and plant community. It's an intelligence that probably deserves to be reinvented for today, nourished by research in science, traditional knowledge and the evocative powers of the arts, so that we may coexist in harmony with the living world around us and in us.

During the territorial explorations of the various species of *Homo* (probably *ergaster*, then *sapiens*), the colonization of new environments led to the spread of forage techniques (shellfish harvesting, fishing, trapping, richer forms of gathering in ecosystems with more differentiated plants), and in the Neolithic led to the revolution constituted by the domestication and storage of food. This diversification of foraging techniques first allowed a release of pressure on tracking skills, which may have become available for other uses. This situation of functional vacancy is characteristic of exaptation. It constitutes the 'liberation' of a trait for an unforeseen change of function (here, of use) likely to revolutionize a form of life. But the matrix form has remained. In the combined animal that is the human being, the module of tracking behaviours is present in the palimpsest (this is the name of those scrolls a thousand times scratched out and rewritten), but subsequent transformations have almost made it unrecognizable.

Our cognitive identity, then, seems to stem in part from the exaptive reserve deposited in us, in the form of behavioural and cerebral aptitudes: these are our sedimented ancestralities. We have mentioned that of the tracker-collector (reading signs, investigation), and that of the social animal dedicated to collaborating and living

The origin of investigation

collectively (theory of mind, abduction of intentions in others). We have also noted, in another register, represented by evolutionary convergences, the patience of the panther, that of the collecting deer which selects its food, that of the bear as tireless taster and that of the dispersing wolf as an explorer of new environments.

We should also mention the fruit- and leaf-eating primate gatherer, which displays a fascination with warm colours, fine memorization skills (seeds hidden by the thousands and found by jays provide an example of the selected major cognitive power of such memory), a power of categorical discrimination based on subtle differences (between a medicinal plant and a poison), capacities for induction (generalizing a property to a whole class of plants, and looking for them among their relatives), perhaps even the use of the concept (in the original form of the naturalists' 'jizz'), and a joyful curiosity towards any new form.

Various elements – framework, cognitive and affective elements – stemming from the life of the hunt, constitute modules that combine with our history as social primates and former fruit-eaters, and with our immemorial status as prey, to end up with the complete and still enigmatic complex that we comprise. These animal ancestralities, the corresponding pressures of selection, the liberation of selected aptitudes towards new uses that escape natural selection, are all the conditions of possibility of what we now call freedom.

The very long history of human hunting was obscured by Neolithic agriculture, which changed our relationship with the search for food. The latter corresponds to only three hundredths of human time, but it has redirected the transformation of mental matrices from hunting life in unheard-of directions, inventing new uses for the hand, for the mind and for desire. The fact remains that the few hundred thousand years of intensive research into animals with a view to our subsistence have probably shaped our inner fabric in depth: *Ecce Homo*, a tracker in a world without prey.

The origin of the existential motive of the quest

If we follow this hypothesis, tracking constitutes a primal and ubiquitous activity of the genus *Homo*, forgotten by every modern human being, that is the basis of part of our cognitive condition. How then could it not also be constitutive of part of our affective matrices?

Temple Grandin, a first-rate specialist in animal behaviour, views the emotional power that feeds our most diverse projects as a derivative of evolutionary tracking. Drawing on advances in animal neurobiology, she translates it into a form of the neuronal love of the very desire to seek, and not the mere pleasure of finding. She thus proposes a theory of the deep meaning of human activity which is the 'quest', mythologized in Western chivalry, Nordic legends, detective novels and probably all adventure literature. Her analysis consists in seeking behavioural peculiarities in animal life that shed light on ourselves.

Grandin draws on the experimental results of neuroscientist Jaak Panksepp. The latter proposes the idea of a 'SEEKING' system (he writes the word in capital letters) to describe the neuronal system as an activity in which the emotions of '*intense interest, engaged curiosity* and *eager anticipation*'[14] are manifested in living beings when they search for food. These emotions are also present during the search for shelter or a sexual partner.

Panksepp's findings are decisive in this regard, because he has linked this circuit to something profoundly new:

> Researchers used to think that this circuit was the brain's *pleasure center*. Sometimes they called it the *reward center*. The main neurotransmitter

The origin of investigation

associated with the SEEKING circuit is dopamine, so they thought dopamine was the 'pleasure' chemical.[15]

It turns out, in fact, that what the laboratory animals under study stimulate for dopamine synthesis is *not* the pleasure system, but the brain's 'seeking' centre:

> What the self-stimulating rats were stimulating was their curiosity/interest/anticipation circuits. *That's* what feels good: being excited about things and intensely interested in what's going on – being what people used to call 'high on life'.[16]

What is true for foraging in general is accentuated for predation in particular, because food is more difficult to obtain in the latter case. We can hypothesize that a predator with a behavioural base inducing intense joy when hunting thus has an adaptive advantage which can be accentuated by selection. More generally, as Darwin saw, evolution tends to inscribe in us pure joys in doing what is good for us (here in the sense of what increases the selective value in one way or another). But Panksepp shows that hunting activates the same networks as the seeking system (curiosity, interest, anticipation) with the same pleasant sensations, the same joys of the quest.[17] It is this thrill of foraging that has been exapted in our daily hunts, separated from any killing and nutritional function.

> Depending on their personalities and interests, humans enjoy any kind of hunt: they like hunting through markets for hidden finds; they like hunting for answers to medical problems on the Internet; they like hunting for the ultimate meaning of life in church or in a philosophy seminar to

The origin of investigation

discover the meaning of our lives. All of those activities come out of the same system in the brain.[18]

The decisive argument that shows the gap between the pleasure circuit and the seeking circuit is the temporality of the activation:

> This part of the brain *starts* firing when the animal sees a sign that food might be nearby but *stops* firing when the animal sees the actual food itself. The SEEKING circuit fires during the *search* for food, not during the final locating or eating of the food. It's the search that feels so good.[19]

Dopamine seems to be the hormone, not of pleasure, but of the quest.[20] Tracking as ardent research is then the animal essence of human quest. The complex set of emotions of interest, anticipation, attentive curiosity, inexhaustible energy, the sensation of 'flow', is an exaptation, therefore a diversion, of a cerebral circuit whose original function was to interest us immoderately in what was important for our survival: following and finding the fleeing animal. 'We remain sharp and alert in the world's extinct forests.'[21]

It is fascinating that the contemporary psychology which investigates the kind of experiences that humans refer to as 'happiness' are of precisely this order. Psychology isolates the idea of 'flow', or optimal experience,[22] and uses it to describe this inner state of intense active concentration, oriented towards a deeply desired goal, this state in which the ego disappears in favour of a tension of the whole being in a quest where all the powers of the individual are called upon. 'Be the leap. Not to be a feast, its epilogue', writes French poet René Char in *Feuillets d'Hypnos* (*The Notebooks of Hypnos*).[23]

The origin of investigation

The exaptation of our seeking system is the salt of our lives: it forms the basis, within the folds of our brains, of our quests, our projects, our vital and mobile strength, our ability to accomplish great things. This behavioural matrix is zoomorphic in the sense that it is through attention to the finesse of a behavioural and emotional complex *in animals* that we end up better understanding who *we* are, by the weaving of evolution that binds us to them. Grandin thus proposes an animal theory of the joys of the quest: Don Quixote is the example of the continuous activity, raised to an incandescent level, of the seeking system of the animal brain.

Grandin is a great ethologist of human beings: she gives an account of who we are by doing justice to the great power of animal life in us, to its plurivocal and subtle determination, far from any simplistic physiochemical determinism. Far from detracting from our humanity, she vitalizes it. She makes visible the mystery of being an animal body, far from the grim jubilation of degrading human beings by reducing them to animals, that characteristic of a certain reductionism which confuses bodies with machines, animalism with primitive baseness. She keeps us from forgetting that we are first and foremost a body in the spiritual experiences of existence: love, fear and existential anxiety, the highest forms of thinking, research and curiosity, desire, peace.

This is the strength of an analysis of human life in biological terms: it manifests itself when the finer and higher aspects of human existence can be made intelligible without degrading them.

Tracking today is taking on another dimension. It is no longer just a folkloric naturalistic practice; it is a question of drawing, in the words of Paul Shepard, 'from the sources of the Pleistocene'.[24] Far from being a romantic experience of living naked in the woods, this connection takes a precise form: it consists in letting a fragment of our human biogram rise up within us, coinciding with the original act of tracking, which actually shaped part of our cognitive and emotional skills, and thus part of who we are. We can then experience what happens when the long human past rises within us and coincides with the present moment.

An origin of public affairs

This philosophically enriched tracking, finally, is not a solitary *robinsonade*: we track often, joyfully, in the company of others. This intensifies the quality of attention and the fabulatory power of tracking, where storytelling allows hypotheses to be made which are then tested on the ground (nothing can exist without leaving traces).

The tracker here is neither a romantic walker, alone in the sublime forest, nor that other romantic figure, the mute Indian who understands the clues from a single frown: he chatters in a low voice, he interprets, he argues. We tell stories, we denaturalize living creatures: we need to historicize them, make them more complex, restore to them the unpredictable texture of their individual and collective existences. Confronted with nothing but traces, those tiny residues that the past offers to the present, deprived of the visible action of beings, deprived of their presence, it is the power of multiple narratives, of embedded hypotheses, which brings ghosts to life, which rematerializes before our eyes those who left those pawprints.

One morning at Urn Lake, Ontario, in Algonquin Park, we find in a stream, deep in a bewitched glade, the trail of a moose. It's impossible to determine its sex, but it is colossal, its hoof is over fifteen centimetres long (the largest individuals are up to two metres at the withers and weigh 600 kilograms). He's heading upstream. At the first bend, we lose his track. We hold council, first in the sign language of tracks, but very quickly we lack the necessary grammar to compare and contrast our versions. Then, in a low voice, one of us defends the idea that he went upriver, because his route was predetermined; the other says he turned towards a grassy flat; the first pouts: it doesn't make sense, this is too convoluted a turn that makes him pass through a clump of pines that's too tight for him; the second shakes his head and points to the area: let's see. There, indeed, are its magnificent traces on the grassy flat, with that characteristic right-right left-left pace, placid in its great elegance. We lose him again when we come out of the flat. The two trackers look at each other, sceptically: the first makes the sign meaning 'speculative tracking'. This means that we stop looking systematically for the tracks, one by one, but that we look up, imagining the animal on its displacement vector as far as the last identified tracks, and that we then mentally project where he must have gone, where the next tracks will be found. And so we head straight there, without looking randomly at the ground.

Very fresh droppings tell us that the animal can't be far away, we shouldn't take it by surprise. The lively debate continues, but in murmured tones, as we squat around the traces we have found. After several fiery tirades, lost and found tracks, we end up isolating the small fir forest where he most likely took refuge for a nap (elks here are nocturnal animals). There are extremely fresh droppings just on the edge of the very dense, impenetrable fir forest. Entering it would only drive him back or make him flee blindly. So we settle down to keep lookout, in a valley, embraced by the stream and the foliage, and we wait with smiles and silence.

When we are tracking in a group, then, there is no meditative silence, but the infinite confabulation, the joy of being together and with creatures other than human beings, and of exhuming possible pasts by telling stories. Not silent human beings contemplating the sublime, but talkative animals investigating the mystery of a shared world. Everything calls for conversation, the sharing of signs, clues, markings, smells, mating display and talkative pheromones – especially since neither they nor their meaning are easily accessible. Living territories are once again becoming a place of commerce in the old sense of the term, a colourful and cosmopolitan trade between plural forms of life – not a place of solitary contemplation.

It is a place of intense sociability, and here we glimpse one of the possible meanings of that strange formula mentioned in the introduction: the Algonquins spontaneously maintain social relations with the forest. Writing about these experiences, then, has nothing to do with 'nature writing' anymore – firstly, because it is *as animals* that we track animals and write these stories; and secondly, because here, there is no 'nature', there are only living territories, inhabited by living beings who are also inhabited, with their histories, their relationships and their forms of modus vivendi, their geopolitics specific to those who live exposed to each other: and none of this is nature in the traditional sense.

If you want to deduce the consequences of this omnipresence of debate in the activity of tracking, there is something considerable in this phenomenon, found over the first hundreds of thousands of years when *Homo sapiens* (and probably his ancestors) practised persistence hunting, intent on not losing the animal's track. It is not through mystical intuitions that we track; it is with an incandescent alloy of body, senses, imagination and argument.

If we dump the clichéd image of the tracker as a solitary and silent Indian who, with nothing but his intuition and without a word, without arguing, with his senses, grasps the hidden essence of things, then there remains the image of a group of trackers, in an immemorial past, around a set of interwoven traces. What are they caught up in? In an interminable chatter, woven into tight arguments, where everyone puts forward their point of view. The anthropologist Louis Liebenberg has clearly shown in his work that contemporary Bushmen trackers perpetuate this practice of *a collective debate in order to interpret tracks* (he also sees in it the premises of the scientific system of peer evaluation of the hypotheses of others).

We can go back further, and reconstruct another story of the origins of this phenomenon: that of the invention of debate to develop a common story out of disagreements. The origin of collective chatter to produce a common version in the face of an opaque and shared situation – and to chart a unified path for the group. This is something like the birth of collective intelligence when tackling common problems; much later, mixed with a thousand other things, it will take the form of 'public affairs', the Agora of Athens. It's *Nuit debout*,[25] but with the pragmatic pressure of hunger, urging people to decide and act.

It was here too, among countless other events, that the human primate began to become who he is now, to take on this strange form of life. The historian Marcel Detienne, in a famous book, ascribes the birth of political reason, as an art of argued, egalitarian and collective dialogue, to the development of a common formulation drawn from the disparate chaos of the multiple experiences of each individual, with the aim of finding a way through, in the tradition of the Athenian army of *pooling* the situation. One example of this is seen in the circle of warriors at the beginning of the *Iliad*, but it was extended, in classical Athens, to all citizen soldiers. Detienne contrasts the masters of truth, the exclusive bearers of a sacred word, with the secularization of utterance made possible by the historic reforms linked to military conscription in Athens: it was in the ranks of this army of

The origin of investigation

citizens that speech as dialogue was born – an 'equal right' to speak in order to 'discuss common affairs'.[26] No more kings and soothsayers holding a monopoly of truth: now, truth is built up together, through egalitarian dialogue between those who are considered as equals (and of course, most people were excluded from this equality – but that's another story).

On its own scale, and as a way of moving from the sacred masters of truth to secular political dialogue in ancient Greece, this hypothesis is probably very correct and very alluring. But if we are looking for an account of the origins of egalitarian argumentative debate for dealing with common things, we should find something that shows a little more seniority and generosity towards other peoples.

It is fascinating to discover how much we seek to base our accounts of the origins of what we are or want to be (here, the account of how democratic speech developed), in such a *recent* past (Greek or Judeo-Christian, essentially), all in order to de-animalize and sanctify it, far removed from our animal origins. And yet hundreds of thousands, indeed millions of years constitute our actual genealogy. Telling the story of human beings starting with the birth of civilizations and of writing around 3000 BC, is like trying to write the biography of a man who died at the age of one hundred years, starting the story at the beginning of his ninety-ninth year – as if nothing that took place between his birth and his ninety-ninth birthday had played a serious part in who he was.

The first outline of collective reason as an art of argued and egalitarian conversation probably did not originate among the Greeks or somewhere in so-called civilized societies during the last ten millennia. It probably arose – this is a more reasonable and more parsimonious hypothesis – gradually, by combining a thousand animal powers acquired over hundreds of thousands of years, partly

The origin of investigation

in the activity of tracking, where the group was meant to find the animal together, by *collectively* interpreting traces right there, in front of everyone's eyes.

This speculation deserves a more careful and rigorous examination than this reverie permits, but it can be summarized as follows: when the group is gathered in front of an animal track, they need to determine which way to go. Everyone evaluates the different versions of what has happened; they compare their respective merits (as the Bushmen of the Kalahari still do today), each person saying what he has to say, each person being listened to according to the quality of his arguments or his demonstrated talents to reconstruct tracks in the mind. Then, together, they can pave the way for action.

This is a practice that requires a collective determination of what is worth pursuing, and in what direction. The argument is woven into the informed imagination, based on things that are unseen, in the service of stories that must be told if the truth is to be told, and to orient ourselves together on this earth.

It's something like the powers behind a whole swathe of public affairs, behind one meaning of the word 'democracy', born in the open air: argued and collective debate aimed, on the basis of individual and segmented perceptions, to make a common story on which we agree, to collect in a converging beam the different versions that everyone hides within their skulls, and allow them to move forward together, for some time, on the same path, on the same track.

Imagine, in an immemorial past, a band of indistinct hominids, endowed with a protolanguage that escapes us, isolating a track in the sand. They stop at the last pawprint found, the point where they lose the trace. One points his spearhead towards the sunset, and another points his spearhead towards the east: he has seen something else in

The origin of investigation

the enigma of the ground, that visible thing which forces one to see the invisible, to imagine and to think. So they gather in a circle, and engage in gestures, phrases, a confabulation that we can't follow, submitting each other's version to the intelligence of the others for assessment, and together determining where we shall go.

Notes

Notes to Preface

1 Jean-Christophe Bailly, *Le Parti pris des animaux* (Paris: Christian Bourgois, 2013).
2 The idea of thinking about the relationships with living things in an inchoate sense is formulated by Baptiste Morizot in an interview with Pierre Charbonnier and Bruno Latour: 'Redécouvrir la terre', *Tracés. Revue de sciences sociales* [online], 33, 2017, posted on 19 September 2017, accessed 14 December 2017. URL: http://journals.openedition.org/traces/7071; DOI: 10.4000/traces.7071.
3 A very good example of this intimacy without proximity can be found in Jacob Metcalf's article on human–grizzly encounters: 'Intimacy without Proximity: Encountering Grizzlies as a Companion Species,' *Environmental Philosophy*, vol. 5, no. 2, autumn 2008.
4 See Morizot, 'Redécouvrir la terre'.
5 My own work is to some extent a response, at once speculative and pragmatic, to the richly suggestive remarks made in Bruno Latour, *Où atterrir? Comment s'orienter en politique* (Paris: La Découverte, 'Cahiers Libres', 2017).
6 Akira Mizubayashi, *Mélodie, chronique d'une passion* (Paris: Gallimard, 'Folio', 2013).
7 Baptiste Morizot, *Les Diplomates. Cohabiter avec les loups sur une autre carte du vivant* (Marseille: Wildproject, 2016), p. 149.

Notes to Preamble

1 I have translated Morizot's French quite literally, even though in English we would not refer to 'going into nature', since he is making a philosophical point about nature as 'other' to culture. (Translator's note.)
2 Philippe Descola, *Beyond Nature and Culture*, translated by Janet Lloyd (Chicago, IL: University of Chicago Press, 2013).
3 Descola, *Beyond Nature and Culture* (translated from the French).
4 Gilles Havard, *Histoire des coureurs de bois, Amérique du Nord 1600–1840* (Paris: Les Indes savantes, 2016). A *coureur des bois* was a French-Canadian trader, usually in fur, who worked closely with the indigenous people of North America, mainly in the seventeenth century. (Translator's note.)
5 Walt Whitman, 'The Song of the Open Road', in *Leaves of Grass* (1855).
6 Emanuele Coccia has written eloquently on this phenomenon in *La Vie des Plantes. Une métaphysique du mélange* (Paris: Rivages, 2016).
7 Claude Lévi-Strauss and Didier Eribon, *Conversations with Claude Lévi-Strauss* (Chicago, IL: University of Chicago Press, 1991), p. 193.
8 Walt Whitman, 'The Song of the Open Road,' in *Leaves of Grass* (1855).

Notes to Chapter One

1 René Char, *Feuillets d'Hypnos* (Paris: Gallimard, 1946).
2 Adolf Portmann, *Animal Forms and Patterns: A Study of the Appearance of Animals* (New York: Schocken Books, 1967), p. 196.
3 Jean-Marc Moriceau and Philippe Madeline (eds.), *Repenser le sauvage grâce au retour du loup. Les sciences humaines interpellées* (Caen: PUC, 2010), p. 117.

4 Aldo Leopold, *A Sand County Almanac* (New York: Oxford University Press, 1949), p. 144.
5 Konrad Lorenz, *Behind the Mirror: A Search for a Natural History of Human Knowledge*, translated by Ronald Taylor (London: Methuen, 1977).

Notes to Chapter Two

1 Edward O. Wilson, *Biophilia: The Human Bond with Other Species* (Cambridge, MA: Harvard University Press, 1984), p. 101.
2 The 'Smokey the Bear Sutra' (1969) is a poem by Gary Snyder.
3 On the relationships between First Peoples and animals as diplomatic relations, see Paul Shepard, 'On Animal Friends', in Stephen R. Kellert and Edward O. Wilson (eds.), *The Biophilia Hypothesis* (Washington, DC: Island Press, 1993).
4 The Marines are another sort of animal, some would say. But it is a mistake to equate animal interactions with physical aggression: diplomatic relations are no less animal than conflicts with fangs and teeth, for ritualized dialogue is no less common than physical confrontation in ecological relationships between rival or commensal animals.
5 Cristina Eisenberg, *The Carnivore Way: Coexisting and Conserving North America's Predators* (Washington, DC: Island Press, 2014), p. 99.
6 David Quammen, *Monster of God: The Man-Eating Predator in the Jungles of History and the Mind* (New York: W. W. Norton & Company, 2004).
7 Val Plumwood, 'Human Vulnerability and the Experience of Being Prey' (1995), *Quadrant*, vol. 39, no. 314, pp. 29–34. (All quotations from Plumwood are translated from the French. Translator's note.)
8 Ibid., p. 31.
9 All the energy that nourishes living beings on Earth comes from the sun

through photosynthesis (apart from the rare bacteria known as chemotrophs at the bottom of the oceans, which use the oxidation of chemical compounds as an initial source of energy).

10 On the cosmology of Siberian shamanism, see Roberte Hamayon, *La Chasse à l'âme. Esquisse d'une théorie du chamanisme sibérien* (Paris: Société d'ethnologie, 1990). See also, on the general scheme of the cosmos as reciprocal predation, Eduardo Viveiros de Castro, *From the Enemy's Point of View: Humanity and Divinity in Amazonian Society* (Chicago, IL: University of Chicago Press, 1992).

11 Plumwood, 'Human Vulnerability', p. 34.

Notes to Chapter Three

1 Paul Shepard, *Nature and Madness* (Athens, GA: University of Georgia Press, 1988), p. 10.
2 Omar Khayyam, *Rubayyat*, quatrain 71.
3 Charles Darwin, *Notebooks (1836–1844)* (Cambridge: Cambridge University Press, 2009), p. 524.
4 We owe this intuition of animal ancestralities to Paul Shepard; see, in particular, his chapter on the eye in *Only World We've Got: A Paul Shepard Reader* (San Francisco, CA: Sierra Club Books, 1996).
5 Saint Augustine, *Oeuvres complètes* (Bar-Le-Duc: Éditions Raulx, 1866), volume XII, chap. 15.
6 Katmai National Park in Alaska has installed cameras that continuously film certain key points of its wilderness. In summer, you can watch grizzly bears fishing for salmon in the rapids of a torrent from your computer screen, live, for days on end, without commentary, without editing, without staging. See www.explore.org/live-cams.
7 Exaptation is a concept in evolutionary theory that describes unforeseen changes in function: biological traits selected for initial use are

subsequently diverted to a new function or use. For example, the feathers of the dinosaurs who were the ancestors of birds were not selected primarily because they allowed flight, but for thermoregulation or display. Only subsequently did they facilitate the onset of flight. See Stephen Jay Gould and Elisabeth S. Vrba, 'Exaptation – A Missing Term in the Science of Form', *Paleobiology*, vol. 8, no. 1, winter 1982, pp. 4–15.

8 Eduardo Viveiros de Castro, *The Relative Native* (Chicago, IL: Hau, 2016), p. 243.

9 Davi Kopenawa and Bruce Albert, *La Chute du ciel* (Paris: Plon, 2010).

Notes to Chapter Four

1 Georges Le Roy, *Lettres sur les Animaux*, Letter II (Oxford: The Voltaire Foundation, 1994), p. 24.

2 Louis Liebenberg, *The Art of Tracking: The Origin of Science* (Claremont, Western Cape: David Philip Publishers, 1990), p. 38. (All quotations from Liebenberg are translated from the French. Translator's note.)

3 Eduardo Viveiros de Castro, *Cannibal Metaphysics. For a Post-Structural Anthropology*, edited and translated by Peter Skafish (Minneapolis, MN: Univocal, 2014), p. 55.

4 Michel de Montaigne, *Essais* (Paris: Gallimard, 2009), Book 1, chap. 39.

5 See Michael Rosenzweig, *Win-Win Ecology: How the Earth's Species Can Survive in the Midst of Human Enterprise* (Oxford: Oxford University Press, 2003), p. 5: 'We can learn how to reconcile our own use of the land with that of many species. Maybe even most of them. If they have access to our fields, our city parks, our schoolyards, our military bases, and yes, even our personal gardens, then they stand a chance. If they live where we live, then they will have what we have. We would then be able to minimize the risk of extinction.' (Translated from the French.)

6 Anna Lowenhaupt Tsing, *Le Champignon de la fin du monde. Sur la possibilité de vivre dans les ruins du capitalisme* (Paris: La Découverte, 2015).
7 Montserrat Suárez-Rodríguez, Isabel López-Rull and Constantino Macías Garcia, 'Incorporation of Cigarette Butts into Nests Reduces Nest Ectoparasite Load in Urban Birds: New Ingredients for an Old Recipe?', *Biology Letters*, vol. 9, no. 1, 2013.

Notes to Chapter Five

1 Roberte Hamayon, *La Chasse à l'âme. Esquisse d'une théorie du chamanisme sibérien* (Paris: Société d'ethnologie, 1990).
2 *Partition rouge, chants et poèmes indiens*, translated by F. Delay and J. Roubaud (Paris: Seuil, 1988), p. 194.
3 The pharmaceutical industry did not synthesize them until the twentieth century (this is the case with aspirin, a form of which exists in the buds of poplar trees and willow bark as salicin, which becomes salicylic acid once metabolized by the liver).
4 Blaise Pascal, *Pensées* (Paris: Le Livre de Poche, 2000), p. 58.

Notes to Chapter Six

1 See the extraordinary BBC footage on the issue in *Life of Mammals* (2002–2003), with commentary by David Attenborough, available online: https://www.youtube.com/watch?v=826HMLoiE_0.
2 Louis Liebenberg, *The Art of Tracking: The Origin of Science* (Claremont, Western Cape: David Philip Publishers, 1990), p. 119. (All quotations from Liebenberg are translated from the French. Translator's note.)
3 Ibid., p. 83.
4 Ibid., p. 57.

5 Personal communication.
6 Note also the existence of random tracking, which takes place when no clues are visible. According to Liebenberg, the Bushmen use divination with leather discs. It is no exaggeration to think that this is an essential part of shamanism: mastering the random, tracing paths in absolute uncertainty. They check the discs to know, before the hunt, which way to go, because there is as yet no information. Two Liebenberg hypotheses: either the discs are interpreted in the light of what they know about the movements of the game they are pursuing. Or they are used to randomly diversify the routes, to respond to the capacities of game to change its habits according to the repeated habits of hunters, and to introduce a salutary unpredictability and renewal. See Liebenberg, *The Art of Tracking*, p. 120.
7 Ibid., p. 112.
8 For Peirce, 'induction designates the testing of hypotheses, whether this ends with a confirmation or a refutation', according to Claudine Tiercelin, who details this scientific method in three stages; see her *C.S. Peirce et le pragmatisme* (Paris: PUF, 1993), pp. 95–8.
9 See Ian Hacking, *The Social Construction of What?* (Cambridge, MA: Harvard University Press, 2000).
10 Liebenberg, *The Art of Tracking*, p. 60.
11 Ibid., p. 39.
12 For the historian Carlo Ginzburg, the reading of clues is also at the origin of the cognitive capacity which consists in going back, starting from apparently negligible experimental facts, to a complex reality which cannot be experienced directly. According to Ginzburg, it is the oldest gesture in the intellectual history of mankind, the gesture of the prehistoric hunter. For millennia, man, through his hunting activity, has learned to reconstruct the shapes and movements of invisible prey from pawprints left in the mud, and droppings. 'He learned to perform

complex mental operations with lightning speed, in a dense thicket or in a clearing full of pitfalls' (Carlo Ginzburg, *Myths, Emblems, Clues*, translated by John and Anne C. Tedeschi (London: Hutchinson Radius, 1990). – Translated from the French). Tracking would thus be the source of semiotics: a behaviour oriented towards the analysis of the individual case, which can only be reconstructed through traces, symptoms, clues. Indeed, it would be at the origin of the 'index paradigm' that characterizes a whole section of modern science according to Ginzburg: medicine, jurisprudence, history, palaeontology ... and criminal investigation.

13 Stephen Jay Gould, *The Structure of Evolutionary Theory* (Cambridge, MA: Harvard University Press, 2002) (translated from the French).

14 Temple Grandin with Catherine Johnson, *Animals in Translation: Using the Mysteries of Autism to Decode Animal Behavior* (San Diego, CA: Harcourt, 2006).

15 Ibid.

16 Ibid.

17 However, during the hunt and then the killing, the neural circuits of anger (activated in intraspecific aggression, or in self-defence) are not activated. The animal is always 'peaceful and quiet' says Grandin (ibid., p. 164). We see how far the reality of predation is from its moral interpretations. The problem is therefore not the hunt centred on the killing, with its hawkish or testosterone-fuelled imaginary, but tracking as a quest, as a search, as an awakening of the senses and of the animal brain.

18 Grandin, ibid.

19 Ibid.

20 We here find, based on biomorphism, a conceptual nuance proposed by Gilles Deleuze: the distinction of value between pleasure and desire. There is a mistake in relating pleasure to an intensification of existence: pleasure punctuates life, it fills and lulls; dopamine, on the other hand,

is the chemical index of desire: it is dopamine that provides the life-giving joys and the intensification of existence that hedonism thinks it is seeking in pleasure, and through this confusion fails to find.

21 Edward O. Wilson, *Biophilia: The Human Bond with Other Species* (Cambridge, MA: Harvard University Press, 1984), p. 132 (translated from the French).

22 Mihaly Csikszentmihalyi, *Flow: The Psychology of Optimal Experience* (New York: Harper, 2008).

23 Aphorism 187 from René Char, *Poèmes et proses choisis* (Paris: NRF Gallimard, 1957), p. 59.

24 Paul Shepard, *Coming Home to the Pleistocene*, edited by Florence R. Shepard (Washington, DC: Island Press/Shearwater Books, 1998).

25 'Nuit debout', literally 'Standing up at night', was a French protest movement that began in March 2016 and lasted for several weeks. Its aims were similar to those of the Occupy movement: to reject the government's austerity measures and target corporate greed. (Translator's note.)

26 'It seemed to me that an understanding of the sociohistorical context might contribute to this genealogy of the idea of truth. During my research on the Pythagoreans, I gleaned signs of a process which set in motion the gradual secularization of speech. The most important sign was to be found in the military assembly since it conferred the equal right to speech on all members of the warrior class, those whose very position allowed them to discuss communal affairs. The hoplite reform, introduced in the city around 650 BC, not only imposed a new type of weaponry and behaviour in battle, but also encouraged the emergence of "equal and similar" soldier-citizens. At this point, dialogue – secular speech that acts on others, that persuades and refers to the affairs of the group – began to gain ground while the efficacious speech conveying truth gradually became obsolete. Through its new function, which

was fundamentally political and related to the *agora*, *logos* – speech and language – became autonomous' (Marcel Detienne, *The Masters of Truth in Archaic Greece*, translated by Janet Lloyd [New York: Zone Books, 1996]), p. 17.

Credits

An initial version of 'The signs of the wolf' was published in the journal *Vacarme*, no. 70, winter 2015, with the title 'Rencontres animales. Voir un loup d'homme à homme'.

A previous version of 'A single bear standing erect' was published in the journal *Billebaude*, no. 9, September 2016 © Billebaude, Glénat/Fondation François Sommer, 2016.

A very brief section of 'The patience of the panther' appeared in *Philosophie magazine*, no. 111, July–August 2017.

A first draft of 'The discreet art of tracking' was published in the journal *Billebaude*, no. 10, June 2017, with the title 'L'art du pistage' © Billebaude, Glénat/Fondation François Sommer, 2017.

'The origin of investigation' is a revised and enlarged version of a chapter of *Diplomates. Cohabiter avec les loups sur une autre carte du vivant* © Wildproject, 2016.